W9-CKU-741

MIRAGE MEN

MIRAGE MEN

An Adventure into Paranoia, Espionage, Psychological Warfare, and UFOs

Mark Pilkington

A Herman Graf Book
Skyhorse Publishing

Skyhorse Publishing books may be purchased in bulk at special discounts for sales promotion, corporate gifts, fund-raising, or educational purposes. Special editions can also be created to specifications. For details, contact the Special Sales Department, Skyhorse Publishing, 555 Eighth Avenue, Suite 903, New York, NY 10018 or info@skyhorsepublishing.com.

www.skyhorsepublishing.com

10 9 8 7 6 5 4 3 2 1

Library of Congress Cataloging-in-Publication Data

Pilkington, Mark. Mirage men : An Adventure into Paranoia, Espionage, Psychological Warfare, and UFOs / Mark Pilkington.
p. cm.
Includes bibliographical references and index.
ISBN 978-1-60239-800-9 (hardcover : alk. paper)
1. Unidentifed flying objects—Sightings and encounters. I. Title.
TL789.3.P55 2010
001.942—dc22

2010018316

Printed in the UK

CONTENTS

ONE

INTO THE FRINGE

'These ships are there and they can be seen by those who look up'
George Adamski, Gray Barker's Book of Adamski
(Saucerian Books, 1965).

'What in fucking hell was that?' blurted Tim, his voice expressing more surprise than fear – a not inappropriate response from someone who had just seen their first UFO.

It was a bright, sunny afternoon in mid-July, 1995. My friend Tim, my then-girlfriend Liz and I were pulled up with a flat tyre alongside Tenaya Lake on Tioga Pass Road, twenty-seven miles from the eastern boundary of Yosemite National Park. I was twenty-two years old, the same age as the car whose front wheel I was in the process of replacing, a battered, sky-blue 1973 Ford Galaxy 500 with a mattress in place of a back seat. The three of us were almost two months into a twelve-week detour around the United States and the car was on its last legs. The mechanics who had most recently looked at the wheezing two-tonne beast had tried to prevent us from driving any further. That was 200 miles ago and the reason, until Tim began yelling, why I'd been lying underneath it clutching a spanner.

Tim loomed above me, fizzing with incredulity. 'What *was* that?'

'I have no idea,' I replied, 'but I saw one just like it about half an hour ago, a few miles down the road.'

I continued loosening wheel nuts as my mind whirled like a top. Whatever it was we'd just seen was identical to the thing I'd spotted further along the road perhaps twenty minutes earlier.

We'd blown a tyre on our way out of Yosemite. To find a new one, Liz and I had hitched a ride to the nearest town, Lee Vining, a small, one-time mining outpost alongside the calcified alien landscape of Mono Lake. Job done, we hopped into a passing two-seater convertible sports car. Liz, in the front, made awkward small talk with its slick-haired driver while I perched on the boot, clutching our reinvigorated wheel with my feet wedged awkwardly into the space behind the driver's seat.

Cool mountain air flowed over us as we re-entered Yosemite on the windy two-lane blacktop. We were rounding a densely wooded bend on the north side of the road when a glint of light among the trees caught my eye. There, perhaps ninety feet away along a straight firebreak path between tall rows of fir trees, lay something entirely unexpected, hovering, or appearing to hover, stationary, about three feet off the ground. It was a perfect, silvery reflective sphere, perhaps eight feet in diameter, like a large, polished Christmas tree ornament. It reminded me of one of the bells suspended over a verdant landscape in René Magritte's enigmatic 'La Voix des Airs': beautiful, serene, uncanny, wrong.

No sooner had I registered what I was looking at than it was gone, lost behind the trees as we sped along the ever-twisting road. Some seconds later we passed another firebreak, converging on the same point as the last one. The thing was still there: mercurial, immobile, strangely perfect. A brief flash, then it was lost among the trees before, finally, another bend, another path and, once again, that damned sphere. I said nothing as my mind rummaged for an explanation. Neither Liz nor the driver seemed to have seen anything unusual and, even if I had known what to say, speech was impossible over the roaring engine and rushing wind.

The sphere and the forest were soon behind us and we returned to our own vehicle, sandwiched between the twinkling lake and a steep-sided rocky hillside. I kept quiet about what I'd seen as I jacked up the car, clambered underneath and began working on the wheel. And that's when Tim began yelling. All I could see were his ankles and feet, but both he and Liz were making excited noises.

'Quick! Get a look at this! What *is* that?!'

I think I knew what I was going to see when I lurched to my feet.

Glinting in the afternoon sunlight, it came gliding purposefully towards us over the lake, bobbing gently as if carried on some viscous current. Although it looked exactly like the sphere I had already seen, it couldn't have been the same one because it was approaching us from the opposite side of the lake, a third of a mile away. It flew right over our heads, perhaps fifty feet up, utterly silent, unhurried yet somehow determined, and slunk out of sight, following the gentle contours of the ridge behind us. The whole thing took less than a minute.

'What *was* that?' Liz spoke for us all. Emptiness filled the airwaves. Brains scrambled for answers. None came.

Back under the car I went, unfastening a few more nuts, hoping to keep a creeping surge of anxiety in check. No chance.

'Holy shit! Here comes another one!' yelled Tim.

Scrabbling back out, I was just in time to see another sphere, identical to the last, ambling over the lake towards us, following exactly the same path as its predecessor. Up it went, over the ridge, as calmly as if it passed by there every day.

I rushed into the car for a camera, but it was too late. The sphere was gone. No more followed.

A strange enough story and a true one, not unlike a thousand other UFO tales. This one is made a little stranger by the fact that I was somewhat interested in UFOs at the time. OK, I'll be honest, I was obsessed with UFOs at the time. I'd been fascinated with the supernatural and the anomalous all my life – I was reading H.G. Wells and Bram Stoker while most kids were still reading Enid Blyton – but somehow, during the late 1980s, UFOs gradually became a primary concern for me.

In 1989, aged sixteen, I had my first sighting, in the south of Spain. A friend and I watched nine glowing orange balls roll along the horizon in a steep sine wave pattern. I remember that they flowed, one after the other, as if moving through thick fluid, connected by an invisible thread. Neither I nor my friend had been particularly astounded by the scene, and alien spacecraft didn't even register on our list of possibilities, but I've often returned to the incident in my mind's eye and wondered: what was it that we saw?

By the early 1990s, UFOs were it for me. In hindsight, I wonder if I was unconsciously caught in a pre-millennial zeitgeist, swept along like many thousands of others in the thrall of the stars. Between the sober, hi-tech thrills of Timothy Good's UFO books, detailing apparently genuine military encounters with seemingly impossible aircraft, and the soul-wrenching alien abduction memoirs of Whitley Strieber, the possibility of alien contact – of life other than our own, of escape from this island Earth – had become all too plausible.

And now I had seen another one.

Compounding the high strangeness factor of our Yosemite incident was the book I had been reading on that leg of the journey, *Into the Fringe* by Karla Turner, a psychologist and UFO researcher who died of a brain tumour a year after our sighting. Her account of her family's UFO experiences is one of the more engrossingly bizarre contributions to an already bizarre field. It also features several appearances by floating silver spheres. Turner describes the spheres as 'repositories . . . where human souls are somehow recycled'. Inside the spheres, human souls that are in some way also alien are implanted into their mothers' wombs, a procedure that is at once surgical and spiritual: a medicalized reflection of the mystical dimension at the heart of the UFO lore.

But there was nothing magical about what we saw that day in Yosemite and, in the years since our encounter, I've often wondered if there might be a mundane explanation for what it was that flew over our heads.

Were they mylar-covered balloons? I couldn't rule it out, though they looked far too solid for balloons, as if they would have emitted a solid clang had we thrown rocks at them (though I'm glad we didn't). The way they bobbed up and down as they flew, like corks on water, and then smoothly followed the contours of the hillside behind us was also most un-balloonlike. Balloons would have bumped awkwardly into the scree-covered slope before being blown over the top. Besides, as I recall, part of the eeriness of the situation was that there was no wind to carry the objects, at least not where we were standing.

Perhaps they were some kind of exotic atmospheric phenomenon, like ball lightning or St Elmo's fire? These electrified gas bubbles are good candidates for some of the more esoteric UFO sightings, and it has been suggested that they might appear silvery in daylight. The US Air Force has been experimenting with generating and steering plasmas as potential weapons for decades; but again, our spheres looked firmly solid, not gaseous.

Were they pilotless drone aircraft of some kind? We weren't far from the China Lake Naval Air Weapons Station, one of the Navy's key testing grounds for new toys, so maybe. But if so, I'd like to know what kind of technology was keeping them airborne. Spherical objects are dropped from military aircraft to train and calibrate radars, but these weren't falling vertically or on parachutes; they were flying, horizontally.

If pragmatism has failed us, how about a mystical explanation? The objects were manifestations that had bubbled up from my own unconscious, inspired by Karla Turner's book, then percolated out into consensus reality – something like the *tulpa* spirit beings of Tibetan mysticism? No? Well it's just a thought, and one I'll confess to having had at the time.

If these were physical objects – and I believe that they were – then without access to the black vault where the US government keeps its secrets, or the bunker that houses its latest military hardware, there's no real chance of finding a satisfactory answer to the question of what we saw that day. And, true to the slippery nature of the phenomenon, reports of similar objects pepper UFO literature at least as far back as the Second World War. For example, a brief *New York Times* item dated 14 December 1944 reads: 'A new German weapon has made its appearance on the western air front, it was disclosed today. Airmen of the American Air Force report that they are encountering silver colored [sic] spheres in the air over German territory.'

Mysterious glowing orbs were seen first by airmen over Europe in 1942. These balls of light, variously described as yellow, orange, silver, green or blue, appeared to follow aircraft, even through dramatic evasive manoeuvres, without ever attacking or damaging them. British pilots referred to the lights as 'the thing', while Americans called them 'foo fighters', after Smokey Stover, a popular cartoon fireman who called himself a 'foo fighter' and had the catchphrase, 'Where there's foo, there's fire!'. Michael Bentine, a former British Air Intelligence officer (and *Goon Show* comedian) debriefed several aircrews about mysterious lights that had harassed them as they flew over the Baltic. The bombers had fired at the lights, which didn't respond: 'These lights didn't seem to do anything, just pulse and go round. We put it down to fatigue, but later, after I had sent the reports in, an American G2 Intelligence Officer told us that their bombers saw lights in the sky – "foo-fighters" he called them.'[1]

Foo fighter reports were taken seriously by the Air Ministry, even if they tended to result in a ribbing from other pilots. Were they an unusual natural phenomenon like ball lightning, or were they, as many speculated, a secret weapon, some new kind of flak or decoy intended to put the frighteners on enemy pilots? Were they radio-controlled? Did they contain a mechanism that allowed them to home in on other aircraft? Bentine recalls debriefing Polish pilots who were pursued by silver-blue balls during a 1943 raid on Peenemünde, where V-2 rockets were being manufactured. Was there a connection to other advanced technologies being developed there? We don't know and surviving wartime records don't provide any answers. Bentine's own conclusion, that if it didn't attack the Poles' aircraft then 'it wasn't a very effective weapon', might seem cavalier, even naive to us today, when decoys and electronic counter-measures (ECM) are a standard part of warfare. But it reflected the attitudes of his superiors, who did not, as far as we know, pursue the question any further. After all, there was a war on.

So what about the spheres I saw in Yosemite? An American with an intelligence background and an interest in UFOs told me that they were US military reconnaissance drones, perhaps lending weight to the China Lake theory. A psychic who claimed to have done 'remote viewing' (RV) work for the US government (psychic spying) told me that the spheres were extraterrestrial in origin and were well known to certain government groups. One US Army colonel told her that the spheres congregated in large numbers somewhere in Kansas, forming geometric shapes across the landscape.

A likely story, you might be thinking: perhaps these things also make crop circles in the vast Kansas prairies? But her account would come back to haunt me some years later as I read *Swamp Gas Times* by the science writer Patrick Huyghe, who recounts a tale told by a Kansas prairie man in the 1980s. The farmer described the pleasures of working the fields on a clear night in his state-of-the-art combine harvester. He loved being out there, he said:

> 'Until they come.'
>
> '. . . Who's they?'
>
> 'When the lights come down,' he said. 'One minute they're not there and the next minute you look out the side window and there they are, keeping pace with you. Then as fast as you can blink,

they'd circle around and be on the other side.'

'UFOs?'

. . . The man wouldn't say what they were, just what they were not: 'Helicopters, aircraft, headlights, reflections . . .'

'So what do you do?'

'I do what I have to do; I just do my job,' he replied.

'Eventually, they just shoot up in the air and disappear. It's really unnerving.'

Maybe the spheres from Kansas were holidaying in Yosemite that day in July 1995, just like we were.

TWO
THE COMING OF THE SAUCERS

'If there is no truth today, there will be myths tomorrow'
Yuli Khariton

SEPTEMBER 2004, CAFE BLISS, DALSTON, LONDON
John Lundberg and I were meeting at our favourite eatery, named after the impact that its take-no-prisoners greasy breakfast had on one's digestive system. John and I met in 1998, when I joined his crew of crop circle makers. Yes, people make crop circles – all of them – and have done since the mid-1970s.

John and his fellow crop-flatteners had been busy every summer since the early 1990s, and they still are. I encountered them as a journalist for *Fortean Times* magazine and ended up joining their team. Although I stuck with it for several years I was never a very good circle maker, but I never tired of working the fields under a night sky (except when it was raining) and was endlessly fascinated by the logical contortions that believers in the crop circle mystery would go through in order not to confront what – to anyone not spiritually, emotionally or financially invested in the phenomenon – appeared to be the bleeding obvious.[1]

On that partcular day, however, we weren't talking about crop circles. John was busy making films and had just wrapped up a short documentary about a crop circle researcher who had got into trouble with the government. He slid into the cubicle seat opposite me, wearing his customary outfit of iPod earphones, army-green puffa jacket and Aphex Twin sweatshirt. Tall and solidly built as his Viking surname would suggest, with his close-cropped hair John might appear intimidating if he wasn't usually smiling. As soon as he'd ordered his vegetarian breakfast we got down to business.

'I've been talking to someone at the CIA,' he said in a hushed voice. 'Everything he's told me so far has turned out to be a lie, but he's a friendly guy and I'm sure he knows a thing or two. At the end of our conversation he said that if I was interested in UFOs I should make a film about someone called Richard Doty. Do you know who this character is?'

I swallowed my baked beans, had a gulp of tea, took a deep breath and began talking. Richard C. Doty was a Mephistophelean character who haunted the underworld of UFO literature. To some, Doty was a dark knight, caught between the intelligence world that he had once operated in, and the world of the UFO researchers to whom he provided incredible information about the alien presence on Earth. To others, he was a pariah, a tool of the government conspiracy, a sower of disinformation and a traitor to the cause of shattering UFO secrecy. In other words, Doty was our kind of guy. The characters who are drawn to liminal phenomena like UFOs and crop circles are an abiding fascination to John and me. These things don't happen in a vacuum – they need people like Doty, and us, to feed them and give them their strange form of life: one that lies somewhere between everyday fact and elaborate fiction. In other words, if a UFO lands in a forest and no one's there to see it, was there ever really a UFO?

In the late 1970s and early 1980s Doty had worked for the US Air Force Office of Special Investigations (AFOSI), which acts as a sort of internal FBI for the Air Force. Most of the time AFOSI, or more usually just OSI, investigates crimes committed on US Air Force bases at home and abroad – anything from petty theft, to drug dealing or murder. AFOSI is also tasked with detecting and deterring any threats to the Air Force and its operations, a counter-intelligence and counter-espionage role that is vitally important in maintaining the technological advantage over its enemies. The US Air Force has been the world's leader in developing new aviation technologies for decades, and AFOSI has played a critical part in this.

As a special agent for AFOSI, stationed at Kirtland Air Force Base in New Mexico, Doty became embroiled in one of the strangest espionage campaigns of the post-war era. It's a story that was never intended to be made public, but it was. Whether the campaign's exposure was Doty's fault or, as some suspect, he was merely the fall guy for a larger operation, it exposed some of the most sensitive machinations of the Air Force to public scrutiny and revealed, for the first time, what many people had always suspected: that the US government was telling lies about UFOs; just not quite in the way that the UFO community wanted to believe.

The story begins in 1979 with Paul Bennewitz, a brilliant engineer and physicist in his early fifties. His company, Thunder Scientific, developed temperature gauges, compasses and other equipment for the Air Force and NASA from workshops on the border of Kirtland. Bennewitz himself lived with his wife and kids in the swanky Four Hills estate on the north side of Kirtland, from where he had a good view of the base and the Manzano mountains in particular, hollowed-out twin peaks that at the time held one of the largest stockpiles of nuclear weapons in the USA.

In July of that year, from his roof deck, Bennewitz began filming strange lights flitting and bouncing around the Manzano area, and recording radio transmissions that he felt were associated with them. Being a responsible citizen, not to mention an Air Force contractor, in 1980 he decided to tell Kirtland Security what was going on. While an undoubtedly brilliant scientist, Bennewitz was also, like many brilliant people, a little on the eccentric side, and had reached the conclusion that the lights could only be highly advanced aircraft piloted by extraterrestrials. He also surmised that their intentions were anything but friendly, and that's what he told the Air Force.

So far, so peculiar, but here's where it gets odder, and very sinister. Bennewitz, who died in 2003, aged seventy-five, was a good man and a true patriot. The Air Force could easily have brushed him off with a 'thanks for looking out for us, these are our own classified aircraft, so we'd rather you just ignored them and didn't tell anyone about what you've seen'. Instead they, or rather AFOSI, decided not just to encourage Bennewitz in his harmless delusions but to amplify them to a volume that would eventually push him over the brink and into madness. For the next few years AFOSI passed him faked government UFO documents, gave him a computer that appeared to be receiving transmissions from the malevolent ETs and created a fake UFO base in remotest New Mexico. All this for one eccentric scientist.

Richard Doty's role was to befriend Paul Bennewitz and steer him deeper into his *War of the Worlds* fantasy. At the same time, Doty was secretly liaising with at least one respected UFO researcher, William Moore, who provided AFOSI with the latest details of ongoing investigations and research in the UFO field. Moore's information was then used to generate bogus government documentation that corroborated the UFO community's suspicions of a top-level UFO cover-up and drew his fellow researchers into a rich pseudo-history of human–alien interaction that stretched back at least two thousand years. Moore, for his part, claimed to have been co-opted with the promise of *genuine* government UFO documents that would prove, once and for all, that extraterrestrials really were visiting planet Earth, and that the US government was sitting on the biggest story in human history.

This twisted pretzel of a campaign lasted until the late 1980s, culminating with the fracturing of both the American UFO community and Paul Bennewitz's mind. Doty's actions were eventually exposed and, after some time working with AFOSI in

West Germany, he retired from the Air Force to become a New Mexico state trooper. And that was as much as I, or anyone else at that time, really knew about Richard Doty.

For me the really interesting part was that Doty and Bennewitz were the conduits, if not the source, for much of the UFO mythology that had emerged since the early 1980s. Stories about crashed UFOs, US government pacts with nasty ETs, alien harvesting of cattle and manipulation of human DNA, which had gained in potency and authenticity as they were retold through countless books, articles, films and TV documentaries. This was the forge of late-twentieth-century folklore, the heart of America's Cold War dreaming and the world in which John and I, with our crop circle work, were already a small part.

I had no idea whether Doty was a maverick or simply one of many agents working the same beat, though America's intelligence agencies had always been associated with the UFO story. Within the UFO community it was assumed that the CIA, the National Security Agency and others were tools in the cover-up of the Truth, but the Bennewitz affair suggested that the opposite might be the case, that these agencies were in fact *responsible* for much of the UFO mythology.

In the early days of the Cold War, America used radio transmitters to broadcast propaganda deep into the Soviet bloc. Every large Russian town ran an 'Interference Activity Service', employing hundreds of 'jammers' who used electronic tones, tape recordings, rattles and voices to block out these hostile American signals. Creating noise, a surplus of information and bogus documentation – data-chaff known in the business as disinformation – is a favourite technique of the intelligence and counter-intelligence agencies. Was this what the Bennewitz affair was really about? If so, then what was the signal that they were masking?

I'd read other stories linking UFOs and espionage: in the early 1950s the CIA smuggled Hungary's crown jewels out of the country disguised as UFO parts, and in 1991 MI6 tried to smear potential UN Secretary General Boutros Boutros Ghali by connecting him to outrageous stories about extraterrestrials.[2] Snippets like these hardly suggested that the intelligence world was desperately trying to keep the lid on the boiling pot of extra-terrestrial reality, more that the UFO was just another toy for them to pull out of the box when it seemed appropriate.

So why, we had to wonder, did John's CIA contact want John to make a film about Richard Doty? It was certainly an intriguing idea, albeit one that we were unlikely to get very far with: Doty had been off the UFO scene for years and the chances of us getting to interview him were next to nil. It didn't help that the UFO scene had itself been stagnant for almost a decade; even the Internet seemed to have broken off its love affair with our alien friends. Interest had peaked in 1997, when *The X Files* was at the height of its popularity and something very large had appeared to drift silently over Phoenix, Arizona that March; but since then there'd been very little zeal for the subject, a trend made clear by the paucity of UFO news clippings being sent in to *Fortean Times*. The only ones I could remember were about the closure of UK UFO organizations and the 'death of ufology'. No, this was not a time to be chasing UFOs, or even stories about UFOs. But why should that stop us?

We decided to make a film about Richard Doty and the Intelligence world's involvement in the UFO community. Perhaps by the time we finished UFOs would be back in vogue again. Stranger things had happened.

Fired up with excitement about our new project, John and I went our separate ways, but as soon as I got home and thought about what I'd just got myself into, my initial enthusiasm began to melt away. I remembered the last time that the world had succumbed to UFOria, when I had found myself, like many others, deeply entrenched in the frontlines of belief. Did I really want to go through all of that again?

UFOS: NORMAL FOR NORFOLK

A dream I had back in 1995, not long after my Yosemite sighting, reflected the intensity of my obsession with UFOs at the time. It was a strange and powerful dream, of the sort that lingers like a stain on the unconscious for years afterwards. In the dream I was invited, like Paddington Bear, to tea with the Queen, Elizabeth II. I was taken to see her in a glowing silvery carriage. I don't remember the outside of the palace where I met her, or even whether it was a building at all, but the interior was a tourist-brochure fantasy of regal chic, every available surface draped with red velvet and ermine and encrusted with jewels and gilt. The Queen was impeccably polite, as was I. We sipped tea from bone china cups and discussed things; I don't remember what. Then it was time to go.

Her Majesty led me to the palace entrance, the threshold of which was filled with glowing yellow-white light. She took my hand to say goodbye and I leaned forward to give her a peck on the cheek, only to be consumed with horror. From my puckerer's perspective I noticed a discoloured patch where the make-up had been rubbed from the royal jowl, and beneath it was cold, grey, leathery, *alien* skin.

Freudians might interpret this as an infantile power fantasy, merged with a representation of my alienation from womankind. Jungians might read it as an encounter with my own *anima*, the goddess-crone within. David Icke, who wrote in earnest about such things a few years after I'd had my dream, would view it as a glimpse at the awful reality of our shape-shifting, blood-drinking, reptilian extraterrestrial overlords. Many in the UFO community would see it as a screen memory disguising a genuine alien abduction experience. Perhaps it was all these things, but in hindsight it was also a sure sign that I was reading too many books about UFOs.

That autumn, back in the UK, I joined the Norfolk UFO Society (NUFOS), based in Norwich where I was living as a student. A couple of months later the group's young founder suffered a marijuana-induced nervous collapse and I was left running the show.

NUFOS meetings took place every two weeks at Norwich's Ferry Boat Inn on the River Wensum. Sometimes they drew in as many as a hundred people, though the hard core was made up of about twenty, among them retired policemen and RAF personnel. Many of the members sought answers to strange experiences of their own, though an upsurge in media stories about UFOs that silly season, mostly surrounding a hoaxed film of aliens being autopsied by the US government, brought plenty of curiosity seekers our way.

As the society's chairman I usually gave the presentations. I spoke about the American 'remote viewing' psychic spying programme, the Face on Mars, what was really going on at Area 51 in the Nevada Desert (I'd reached the base perimeter on the US road trip), whether the world would end in December 2012 and other now well-worn and rusty staples of the UFO lore that, in pre-Internet days, were still considered fairly esoteric, certainly in Norwich.

NUFOS also carried out investigations, occasionally getting a mention in the local papers. One evening I went to see a man who had filmed an oddly shaped bright light in the sky. It turned out to be the planet Venus – the innocent party in so many UFO sightings – given unusually angular form by the internal shutter mechanism of his video camera. Another time the local paper ran stills from a video showing an orange light drifting through the sky. It was classic UFO footage, a formless, illuminated blob of colour rising against a dark night backdrop. It could have been anything. My phone number was listed in the news report about the film, resulting in a handful of anxious calls about other strange lights in the sky. My standard response was to suggest that the witness kept watching the light until they became too cold or too bored to continue. Then they were to go back outside at the same time the following night: if the light was still there then they didn't need to call me back. No one did.

It was a simple enough equation. UFO reports in the media encouraged curious people to look up at the sky, something most people rarely do. There they would see things they had never seen before and wonder if they'd seen a UFO. It had happened to me on numerous occasions. The most common culprits were bright stars and planets (particularly Sirius and Venus), satellites, shooting stars and aircraft making their descent to land, their front lights appearing to hang motionless in the air. These kinds of sightings make up almost all UFO reports, and they always will. But there are other kinds of UFO reports and we got a few of them, too.

Most spectacular were the flying triangles, variations on which are still seen all over the world. A famous sighting had taken place from an oil rig off the Norfolk coast in 1989 when a mysterious, black, triangular-shaped craft, which became known as the North Sea Delta, was spotted being refuelled by an American KC-135 tanker plane accompanied by two F-111 fighter jets. This was no misidentification of a kite or a seagull: the witness, Chris Gibson, was a former Royal Observer Corps member and knew his aircraft.

These flying triangles – almost certainly advanced military aircraft, or a range of different aircraft – are recurring characters in the UFO story, where they go under names like Aurora, Black Manta or TR-3B. They've been reported at sizes ranging from the size of three (American) football pitches (more than 1,000 feet), to that of an ordinary military plane, and have been clocked at speeds ranging from a hover to faster than the blink of an eye. They are often imbued with special powers like silent flight, invisibility and the ability to defy gravity.

NUFOS had received reports of black triangles hovering over canal boats on the Norfolk Broads and an alarming account from a family who had been shadowed by one on a motorway. It flew so close, they said, that they could have poked it with a broom handle, should they have had one in the car. These reports were intriguing, certainly more exciting than the usual lights in the sky, but what were we supposed to do with them? We could talk to the local RAF, who would suggest that we made a formal report to the Ministry of Defence, but the MOD were hardly likely to give us a spotter's badge for spying on their most secret aircraft.

'What are we going to do about it?' was a regular refrain at our meetings. I certainly didn't want to do what the leader of the Lancashire UFO Society had suggested when his own group were faced with similar flying triangle reports; he'd advocated breaking into RAF Wharton, where the enigmatic aircraft were alleged to be stationed. While it was an interesting idea, I suspected that most NUFOS members would have had trouble climbing the stairs at any of the numerous military installations in our own area, let alone scaling the fence.

Some of our regulars had quite complex and personal relationships with the UFO phenomenon. One woman associated sightings of red balls of light over her home with the chronic fatigue syndrome (CFS) from which she suffered. Another, older, woman was trapped in a wheelchair with an illness so extraordinary that no doctor could diagnose it; she was convinced it had something to do with aliens. When we first spoke she told me that she had seen aliens disguised as logs in her woodpile. She had also seen red lights over her home, like the woman with CFS. Over time, I came to realize that in the two weeks between each NUFOS meeting, the woman would experience whatever had been discussed by the rest of us at the previous gathering. I secretly suspected that she was a model case of what psychologists, publicly at least, call a fantasy-prone personality.

Then there were the psychics; the UFO phenomenon has always drawn them in. As far back as 1945 – two years before the world first heard about flying saucers – American psychic Meade Layne had founded the Borderland Sciences Research Foundation Organization and begun channelling extraterrestrials known as the Etherians. By the mid-1950s clear lines had been drawn between the 'nuts and bolts' ufologists, who tended towards science-orientated research and often came from professional scientific backgrounds, and the channellers and contactees, who were more spiritually inclined. At NUFOS we had them all.

One of our members, Sean, was a scrawny, unkempt man with black bags under his eyes and a permanent thousand-year stare.

He'd done psychic work for the UK government, he told me, and was part of a secret cell whose role it was to combat the Black Magic operations of the Conservative Party. It was obvious, Sean once told me over a Guinness, that Norman Tebbit was the leader of this Tory satanic cabal: 'You only have to look at his eyes.' Several years later, long after my tenure as chairman, Sean was taken to court by NUFOS, accused of stealing funds and property.

George and his wife Janet came late to the group. A friendly couple who exuded an air of utter seriousness, George and Janet channelled messages from extraterrestrials and were shown visions of the future on their television set. One of these visions revealed a mass alien landing on Earth, heralding the apocalypse in the year 2000 (and that was a long time before anyone had even heard of the millennium bug).

At NUFOS, being unusual was one thing, but being uncertain was another. One day, in the early summer of 1996, a committee meeting was called and we assembled in the back room of the Ferry Boat Inn. It had been decided, I was informed, that I wasn't the right leader for the group. I was friendly, young and reasonably bright, but I didn't have answers. In fact my presentations always resulted in more questions, and that wasn't really helping anyone to get to the bottom of the UFO mystery. NUFOS needed a leader who could instil discipline into the group, give it direction and, yes, provide those answers.

So they'd decided that George, the psychic who received alien messages through his television, should be the group's new Chairman. To say I felt betrayed would be overstating it – I had finished my studies and was preparing to move to London in the next couple of months – but I did fear for the group and where it might be heading. I imagined them huddled together on a cold Norfolk night on the Broads, lit only by the flicker of a portable television set, waiting to be transported up on to a passing flying triangle.

In hindsight, I wasn't the best man for the job of running NUFOS. What I thought of as the ideal sceptical middle ground – neither believing nor disbelieving in a theory until the facts made it impossible for you no longer to take a position, and then questioning that too – was impossible to maintain within this kind of community. Anyway, what had I been doing there? Was I really running a UFO society or was I an awkward, even dishonest, participant-observer who had got in over my head? I still don't know, but soon afterwards I moved to London, taking with me a cartoon of an alien being beamed down to the car park of the Ferry Boat Inn, drawn by Harry, an ex-policeman in the group.

Things changed for me after that, as did my relationship to the UFO phenomenon. The skies over London, crowded out by streetlights and tower blocks, filled with mundane air traffic, are rarely as rewarding as those over Norfolk's flat and empty expanses. Over the ensuing years I stopped looking up and my interest in the subject began to fade, as it seemed, did the rest of the world's. I still enjoyed reading UFO tales – the more bizarre the better – and kept an ear on the UFO grapevine for any dramatic new developments in the lore, but there were none that left any kind of impression on me, or, it seemed, the UFO field.

I was growing tired of the relentless crowing from the UFO community that 'the truth was out there', that the greatest event in human history, an exchange with extraterrestrial life and culture, had been consigned to a vault or hangar somewhere on an American military base. If this event had taken place, then where were the signs? Where were the sudden deviations from the script of history? Where were the ET technologies? Who would really benefit from keeping them secret? Who could possibly have more money and power than Madonna, Steven Spielberg or any number of oil-rich sheikhs or oligarchs? If some secret organization did hold the Truth, as the conspiracists believed, what could they do with it that these people couldn't?

The pieces just didn't fit, the evidence wasn't there and the long-promised revelations just weren't coming. It was clear that even if there were real UFOs, then nobody – not the world's governments (secret or otherwise) and certainly not the ufologists – knew very much about them. To paraphrase one ufological commentator, UFO researchers knew everything about UFOs except what they are, why they are here, where they come from and who's steering them. Any sense that I had once felt of a possible non-human presence, an alien sentience behind the phenomenon, had dwindled to almost nothing. That people were still seeing UFOs I had no doubt – they always have and they always will – but I came to feel that the most important part of most UFO sightings was what went on inside the witness, not outside.

I eventually lost touch with NUFOS, but it still exists. The group, and many others like it, serves as a perfect microcosm of the worldwide UFO community, and indeed of any community centred on a mystery. In fact, with their eternal quest for truth and meaning, permanently hampered by the struggles-in-miniature of day-to-day existence and the slow throttling of the imagination by the overwhelming bureaucracy that governs them, these groups probably serve as a decent metaphor for life itself. Would civilized life on another planet really be so different?

THREE
UFO 101

'Then did the people lament and stretch out their hands in despair to the skies. Uncle Prudent and his colleague carried away in a flying machine, and no one able to deliver them!'

Jules Verne, Robur the Conqueror *(1886)*

So I would be chasing UFOs again, a decade older and, I hoped, a little bit wiser than the last time. But where were John and I to start looking for them? Nobody seemed interested in flying saucers, triangles or rhomboids – I wasn't even sure if *I* was still interested – and besides, the world had more pressing matters to concern itself with: there was a war on.

By 2004, contrary to the message presented by George Bush's 'mission accomplished' stunt of the previous summer, the war in Iraq had revealed itself to be far from over. All eyes turned to the Middle East as the situation continued to unravel – but were the watchers all human? Beginning in April, Iran underwent a dramatic wave of UFO sightings, kicking off with a bright disc seen over Tehran and a sphere with two 'arms' over the northern city of Bilesavar. Both were filmed and broadcast on state television. UFOs were spotted flying over Iran's nuclear reactors, also the target of increasingly aggressive US rhetoric. Rumours spread like wildfire and the UFO wave soon picked up strength. Did the lights belong to ET observers or were they American or Israeli reconnaissance aircraft spying on Iran's burgeoning nuclear facilities?

By December, as America's anti-Iranian language hotted up, so did the UFO reports. An Iranian Air Force spokesman detailed sightings over Bushehr and Isfahan provinces, both of which contained nuclear facilities, while a bright shining object was seen hovering over Natanz, home of the nation's uranium-enrichment plant. 'All anti-aircraft units and jet fighters have been ordered to shoot down the flying objects in Iran's airspace,' he warned. According to press reports, the rash of sightings led to the convening of a military and scientific panel to study the UFO problem, though no more news was forthcoming.

Tehran had good reason to be concerned, as this wasn't its first taste of UFO intrigue. In September 1976 an unexplained light was pursued over the city by two Iranian Air Force F-4 fighter jets. As the aircraft approached the UFO their communications were scrambled, as were those of a passing civilian passenger aircraft. The incident, documented both visually and on radar, remains one of the more puzzling cases on record, though there's never been anything to suggest that the light was extraterrestrial in origin.

Nor did I see ETs snooping over present-day Iran; instead the situation brought to mind a scenario that had almost been played out over Libya in 1986. At that time, the CIA, the US State Department and the National Security Council had developed a strategy, codenamed VECTOR, to overthrow the country's irksome Gaddafi regime by convincing the dictator that a major American-backed coup was imminent. A key part of the plan was to fly phantom aircraft over the country using fake radar returns and radio transmissions: when the Libyan Air Force scrambled its interceptors they would find nothing – leading, it was hoped, to high-level bamboozlement and insecurity. The idea was that these 'UFOs', used alongside other destabilizing strategies, would set the needles of paranoia pricking at Gaddafi and his administration, weakening it and creating an environment ripe for regime change. VECTOR was abandoned after the project was leaked to the American press, but is typical of the kinds of games that the US Intelligence agencies play with enemy nations.[1]

While VECTOR never happened, the mysterious aircraft over Iran sounded as if they could have been part of a similar destabilization operation. Whoever was observing Iran's reactors – whether alien, American or Israeli – would have the technology to do so without drawing attention to themselves, either using satellites or reconnaissance aircraft. Whatever was being flown over Iran was intended to be seen, and intended to cause UFO stories to spread. Small ripples of irrationality can make big waves. This is a fact the Americans learned during their own first waves of UFO sightings in the late 1940s, sightings that, curiously enough, occurred prominently around the centres of America's own fledgling atomic programme, places such as Los Alamos and Roswell in New Mexico and Oak Ridge in Tennessee.

FROM ALBATROSS TO ZEPPELIN

We have always been exposed to the heavens, and have always told stories about them. By the time that humans had taken to the skies, they were already crowded with dragons, serpents, ships and armies. Tales of these sky beings are as old as we are, while aerial apparitions had been documented in paintings and pamphlets since the Middle Ages as portents, projections and reflections of human affairs on the ground.

The first outbreak of what we would recognize as modern UFOria was a rash of mysterious airship sightings over Canada, America and then Europe in the last years of the nineteenth century. These large, cigar-shaped dirigibles, many fitted with dazzling headlights, came straight out of Jules Verne's 1886 science-fiction novel *Robur the Conqueror*, in which a maverick inventor travels the world in a propeller-driven, flying airship, the *Albatross*.

As would be the case throughout the twentieth century, these mystery craft clad themselves in the futuristic, fictional aesthetic of their day, always keeping just a few paces ahead of contemporary aviation technology. Some of the news stories were probably the work of impish local journalists cashing in on the airship fad and pranking their readers – Edgar Allan Poe had concocted a fake story of an Atlantic balloon crossing for the *New York Sun* in 1844. Others were likely misidentifications of natural phenomena, transformed into something more spectacular by the excitement in the air. Some, however, seem to have an undeniable ring of truth to them.

An early account on 12 July 1891 is typical of the wave. Residents of Theodore St in Ottawa, Ontario, were startled by the sight of 'a huge cigar, at one end of which there appeared to be a revolving fan, while the other end was enlarged, from which a bright light was plainly visible'.[2] A later report, from the Illinois *Quincy Morning Whig* for 11 April 1897, describes what sounds like a modern-day aircraft, complete with navigation lights on the correct sides of the vessel:

> Men who saw the thing describe it as a long, slender body shaped like a cigar, and made of some bright metal, perhaps aluminum [sic] . . . On either side of the hull extending outwards and upwards were what appeared to be wings, and above the hull could be seen the misty outlines of some sort of superstructure . . . At the front end of the thing was a headlight . . . About midway of the hull were small lights, a green light on the starboard or right hand side, and a red light on the port or left hand side'.[3]

Hundreds of similar reports followed from all over the US, some involving encounters with the airships' pilots, usually portrayed as inventors or military men. A curious footnote to the mystery was found at a Houston junk shop in 1969, a series of notebooks belonging to a German immigrant, Charles Dellschau, who moved to America in the mid-nineteenth century and died in Houston in

1923. The thirteen notebooks are filled with detailed, childish illustrations of fantastical airships, presented as technical drawings for a group of wealthy inventors and aviators, the Sonora Aero Club. Worked in among Dellschau's colourful, whimsical drawings are press cuttings about early aviation experiments. The notebooks have been snapped up by art collectors as early examples of 'outsider' art, but were they more than that? In a previsioning of tales surrounding the early flying saucer sightings, it has been suggested that the Sonora aeronauts were secretly building Jules Verne-esque airships for the nineteenth-century American military. Looking at Dellschau's charming, Monty Pythonesque confections, however, it's hard to see them as anything more than visionary flights of fancy.

Whoever was behind the airships maintained an eerie silence, despite the clamour for answers from the press and public: yet soon history would catch up with them. By July 1900 Count Ferdinand von Zeppelin was test-flying his first airships over Europe. Over the next two decades flight, no longer the sole preserve of gods, dragons and mysterious aeronauts, began to lose some of its mystery. The First World War saw the use of balloons and planes in reconnaissance and dogfights but, perhaps tellingly, there are no reports of anything we would recognize as UFOs. The Second World War, however, was very different, and Michael Bentine's foo fighters weren't the only mystery aircraft taking to the skies.

GHOST ROCKETS

Early on 25 February 1942, less than three months after the devastating Japanese raid on Pearl Harbor and America's entry into the war, something triggered a massive assault by anti-aircraft batteries along the Los Angeles coast. The 37th Coast Artillery Brigade, on edge after a Japanese submarine attack on Santa Barbara the previous day, fired about 1,400 shells into the air, but other than the gunners' own shells, which caused six deaths, nothing came down. A number of unidentified aircraft were spotted in the region that night, but the Japanese insist to this day that they didn't attack Los Angeles. So what prompted the barrage? Was it a stray balloon, an alien craft or just war nerves guiding itchy trigger fingers?

Another puzzling incident occurred on 28 November that year, this time over Turin, Italy. According to official Air Ministry records, the entire seven-man crew of a Lancaster bomber saw a 300-foot-long object cruise below them at 500 m.p.h. It had four pairs of red lights along its length and appeared to leave no exhaust trail. The Lancaster's captain claimed to have seen a similar craft over Amsterdam three months previously.[4] Both these cases and the multiple foo fighter accounts remain genuine UFO mysteries, but it is the battle of Los Angeles that most concerns us, a perfect example of the chaos that, a decade later, America's Cold War guardians feared could ensue in a nation gripped by UFOria.

The UFO phenomenon as we know it really began a year after the end of the Second World War. The *Daily Telegraph* of 12 July 1946 tells the story:

> For some weeks a fair number of 'ghost rockets' going from south-east to north-west have been reported from various parts of the eastern coast of Sweden. Eye witnesses say that they look like glowing balls and are followed by a tail of smoke more or less visible. So many reports cannot be put down to pure imagination in the matter. As there is no definite evidence that the phenomena are of meteoric origin, there is growing suspicion that they are a new kind of radio-controlled V-weapon on which experiments are carried out.

A few days later, the *Daily Mail* reported similar sightings over the Alps. The *Mail* suggested that the rockets were being tested by the Soviets, based on technology captured from Wernher von Braun's Peenemünde V-2 plant on Germany's north-east coast, now in Russian hands. In late July, the British sent two of their most trusted rocketry experts on a secret mission to Sweden. They found that eyewitness descriptions of the objects rarely corroborated each other: some were balls of fire, others missile-like projectiles; some made noise, others were silent; a few crashed into the ground or splashed into lakes; one 'vanished' into the sea at a depth of only three feet.

International concern continued to mount. In late August two American rocketry experts went 'holidaying' in Sweden, and by autumn the mystery had only deepened. Swedish and British officials couldn't decide whether the mystery objects were machines or meteors, though the press had no such doubts. On 3 September 1946 the *Daily Mail*'s top war correspondent, Alexander Clifford, declared that 'The Russians, with tightly sealed lips, are experimenting publicly with a machine that leaves no trace whatever and apparently defies several scientific laws.'

The press had themselves a classic military mystery, but British investigators were tiring of the ghost rockets, especially now that they had finally got their hands on some evidence. A photograph of a flaming object falling to earth over Sweden looked suspiciously like a meteorite, while an alleged rocket fragment, once analysed, was discovered to be an ordinary lump of coke. A top-secret telegram from the Foreign Office on 16 September bristled with impatience:

> We are not convinced that there have been any missiles over Scandinavian territory. . . . A very high proportion of all observations are accounted for by just two meteors visible, one by day and one at sunset, in Sweden on 9 July and 11 August respectively. The residue of observations are random in time, place and country, and can not unreasonably be attributed to fireworks, swans, aircraft, lightning etc. and imagination. Such mass delusions are in our experience not unusual in times of public excitement.'

While the men from the ministry may have put the ghost rocket scare down to 'public excitement', we'd be wrong to underestimate the anxieties raised by the scenario on both sides of the Atlantic. Fears of a Soviet super weapon would only escalate in the ensuing years, becoming the first *de facto* explanation for the flying saucers among the military and the public that would soon plague the US. Whether the rockets existed or not, they were the ghosts of things to come, the opening salvo in the coming Cold War.

America's own obsession with flying saucers began on 24 June 1947, when Kenneth Arnold, a fire-control system salesman and pilot from Boise, Idaho, saw nine fast-flying objects near Mount Rainier in Washington State. Arnold's epoch-making sighting came a year on from Churchill's Iron Curtain speech, and by now the new battle lines were starkly clear. After barely two years of peace, the spectre of total war once again hung heavy over the world and, following the horrors of Hiroshima and Nagasaki, those in power knew full well that another global conflict could be humanity's last.

With the benefit of hindsight we can see in Arnold's encounter, and the media storm it evoked, an inevitable reflection of a society moving so fast that historians and futurologists were left struggling to keep up. That year, 1947, was the one in which a man, Chuck Yeager, flew faster than the speed of sound. It was also the year that ENIAC, the world's first digital computer, was switched on; it was the year of the transistor, the microwave oven, the stereoscopic camera and the AK-47. It was the year that the US Air Force was established as a separate military service; it was the year that the Office of Strategic Services (OSS) was transformed into the Central Intelligence Agency (CIA) and the Truman Doctrine and the Voice of America became the Cold War's first acts of ontological aggression. It was the year that America got its first real glimpse of its own future, and that future was saucer-shaped.

What Arnold actually saw remains a matter of fierce debate within the UFO community. The pilot himself expressed some uncertainty, at different times describing the objects as disc-shaped, heel-shaped and half-moon-shaped, then later drawing one that was crescent-shaped. But what they looked like, even what they were, would end up being less important than their impact on the American imagination. It didn't matter whether they were a squadron of alien spacecraft, a mirage, a flock of pelicans, guided missiles, secret US or Soviet aircraft or captured German Second World War technology. Thanks to a neat turn of phrase from Oregon newsman Bill Bequette, they were flying saucers, and the nation was sent spinning into a state of uncontrolled UFOria that influenced everything from interior design to its own military aircraft.

Within five years, the flying saucer would be imprinted on to the minds of every American who had ever flicked through a magazine, read a newspaper or been to the movies. Saucers dominated the skies, inspired the first frisbee (the Pipco Flyin'-Saucer, launched in 1948), provided material for countless singers, comedians and artists and, alongside James Dean, Elvis Presley and Marilyn Monroe, led the vanguard of Americana over planet Earth.

WATCH THE SKIES!

The first decade of the UFO era is its most remarkable. All the themes present in today's UFO lore were introduced in that first crucial decade: sightings of mysterious craft flying at incredible speeds and performing impossible manoeuvres; the recovery of crashed saucers and their dead alien occupants; contact with flying saucer pilots; terrifying abductions, bizarre experiments and, most importantly, the cover-up of all of this by a government afraid to tell its people the truth. But if for most of the world flying saucers (they became UFOs only in 1952) were a source of wonder and, perhaps, entertainment, for the newly created US Air Force and the intelligence services they were just one big headache.

News of Kenneth Arnold's June 1947 sighting shot quickly around the world. Arnold enjoyed his role as flying saucer ambassador and spoke regularly about his encounter, expressing his belief that the objects were secret aircraft – hopefully American rather than Soviet – and may have been powered by atomic energy. His statements brought him to the attention of both the military and a science-fiction magazine editor called Ray Palmer. Thanks to both of them, Arnold would become not just the world's first flying saucer witness, but also its first investigator – the first ufologist.

Ray Palmer had been publishing tales of visitors from the Inner Earth and Outer Space since taking over as editor of *Amazing Stories* in 1938. At the time of Arnold's sighting, Palmer's magazine had been enjoying unprecedented success. This was mainly thanks to 'I Remember Lemuria', published in 1945, and similarly lurid, 100 per cent true tales about a subterranean, technologically advanced, alien menace called the Dero.[5] The Dero stories were astoundingly popular, boosting *Amazing Stories'* circulation to 250,000 copies a month and making an unlikely star of their author, Richard Shaver, a paranoid schizophrenic welder and painter from Wisconsin, who worked at the Ford Motor Company.

Palmer had also helped give birth to the flying saucer. In September 1946 *Amazing Stories* ran four short items by the science writer W.C. Hefferlin, one of which, 'Circle-winged Plane', described an improbably advanced aircraft being flown over San Francisco in 1927. This prototype flying saucer was powered by the mysterious Ghyt (Gas Hydraulic Turbine) motor and could zip along at a staggering 1,000 m.p.h. (official aviation history has it that Chuck Yeager broke the sound barrier in October 1947, reaching about 887 m.p.h.). Piloted from a dome at the centre of its circular wing, this 'pilot's dream' could reach an altitude of 60,000 feet (officially the U-2 would do this in the mid-1950s) 'with the ability of an antelope to move in any direction at will'. The same issue of *Amazing Stories* featured a terrifying tale by Richard Shaver of alien kidnapping. Nobody reading *Amazing Stories* in September 1946 could have known that the circle-winged plane, renamed the flying saucer, would be a reality within a year, or that Shaver's nightmare vision of abduction by aliens would happen for real a decade later.

The first UFO investigation, conducted by Kenneth Arnold just a month after his own sighting, was instigated through the pages of

Ray Palmer's magazine. It began just days after Arnold's fateful flight, when Palmer received a package containing a letter and some rock-like material. The letter, from Harold Dahl, described an incident that had taken place on 21 June, three days before Arnold's own sighting, close to Maury Island in the Puget Sound, near Tacoma, Washington. A number of mysterious flying 'doughnuts' had passed over Dahl's head, one of them ejecting great gushes of black material, reminiscent of slag or lava. This was what Palmer now held in his hand – actual debris from a flying saucer!

Dahl, a harbour patrolman, was steering his boat near the uninhabited Maury Island, about three miles from shore, when he, his son and his crew saw five flying doughnuts silently circling a sixth that seemed to be in trouble. Dahl described the craft as 'balloons', round but slightly squashed on top. They were about 100 feet in diameter, with 25-foot holes in their centre, hence the doughnut shape. Portholes around the exterior of the craft glowed 'like a Buick dashboard'. As the crew watched, the troubled central doughnut dropped to about 500 feet and, with a 'dull thud', released a large amount of paper-like metal material and molten, black rock. Some of the rock hit their dog, killing it; some burned Dahl's son's arm. Limping back to shore, Dahl immediately told his story to Fred Crisman, whom he referred to in the letter as his 'superior officer', and then went home to recover.

The following morning a man dressed in black appeared at Dahl's house in a 1947 Buick and invited him to breakfast at a nearby diner. Here, the strange man – the prototype Man in Black, a staple of 1950s and 1960s UFO lore – proceeded to repeat every detail of Dahl's encounter and warned him not to talk to anybody else about it. Although perturbed by the situation, Dahl ignored the man's warning and wrote to Ray Palmer, including samples of doughnut slag collected by Fred Crisman the day after Dahl's encounter.[6] Palmer was initially dubious about Dahl's story concerning 'the one that got away', but later changed his mind and, in mid-July, suggested to Kenneth Arnold that as the man who discovered flying saucers, he was the ideal person to investigate the case. He offered Arnold a $200 sweetener (about $2,000 in current value) to help him make up his mind.

On 25 July 1947 Arnold was visited at home in Boise by two Army Air Force Intelligence agents, Frank Brown and William Davidson, who had been tasked with finding out what was behind the flying saucer stories. It was a friendly meeting: the AAF men quizzed Arnold about his sighting and showed an interest in Palmer's invitation, which Arnold had not yet accepted. The agents also spoke to some of Arnold's friends and aviation contacts, expressing concern about his talk of atomic aircraft. One of them was David Johnson, aviation editor of the *Idaho Statesman*. Johnson had himself seen a flying saucer and it was he who finally convinced Arnold to investigate Dahl's claims, promising that his newspaper would cover expenses.[7]

On 29 July Arnold flew to Tacoma, having another UFO sighting en route, this time of around two dozen brass-coloured, duck-like objects flying towards him at immense speed. On landing at Tacoma airfield, Arnold found a room already booked for him at the upmarket Winthrop Hotel. This was strange – only Johnson knew that he was planning to be there that day. Arnold took the room and located Dahl, who drove him to the home of a secretary who was looking after the black slag from the saucer, some of which was being used as an ashtray. The UFO debris looked like ordinary lava to Arnold, but Dahl insisted it was the same material that had struck his boat. Fred Crisman showed up a little later, striking Arnold as an impressive, confident character, in contrast to Dahl, who was rather timid and slow-witted, tending to retreat from conversation when Crisman was around.

Feeling a little out of his depth, Arnold decided to call in E.J. Smith, another pilot who had seen the saucers, to help him with his investigation. The following day, speaking alone to Smith, Crisman explained that the doughnut-shaped craft seen by Dahl were very different to the Mount Rainier objects described by Arnold. He also insisted that neither craft was anything you'd expect the American military to own, mentioning rumours that the Nazis had built flying discs at the end of the war. Was Crisman trying to steer Smith into convincing Arnold that the craft he'd seen weren't American?

Things soon began to get very strange for Arnold and Smith. A call from a United Press International journalist, relating what the two pilots had been talking about in their hotel room, led them to believe that the room was bugged; they searched but found nothing suspicious. Fearing a set-up, on 31 July, Arnold called in Davidson and Brown, the two Air Force Intelligence agents who had visited him in Idaho. For some reason Brown refused to accept Arnold's phone call on his base line and called back from a pay phone, agreeing that he and Davidson would fly up from Hamilton, California that afternoon. Minutes later Arnold got another call from the UPI journalist. He'd had another tip-off, this time from a pay phone, and knew that the Air Force were planning to investigate. Was Brown the informant? No sooner had this call ended when another journalist rang Arnold from the lobby. He, too, knew what was happening and wanted a scoop. Who was passing them information about the investigation?

Brown and Davidson showed up that night and were handed a Kellogg's Corn Flakes box full of debris by Crisman, which Arnold thought was noticeably different from the pieces he'd seen previously. Arnold tried to tell the agents about the hotel room being bugged, but they appeared uninterested, hinting to Arnold that the whole thing was a hoax conducted by the harbour patrolmen and that he shouldn't take it too seriously.

Early the next morning, 1 August 1947, the two agents boarded a B-25 with the box of doughnut debris to take back to California. Shortly after take-off, at 1.30 a.m., a fire broke out in the plane's left engine and it crashed near Kelso-Longview in Washington State. Brown and Davidson both died while two other military passengers survived. One of the survivors described how Davidson opted to stay on board to protect his cargo, while Brown's escape was blocked by a broken wing. The first of August was Air Force Day, when the United States Air Force gained its independence from the Army, making William Davidson and Frank Brown its first casualties.

It was all too much for Arnold and Smith. Distraught over the agent's deaths and frustrated with the lack of a resolution to their investigation, they decided to go home. Before doing so they met Air Force Major George Sander, who requested that they give him all the remaining pieces of doughnut wreckage. Reluctantly, they did so. Sander then drove them to a peninsula near Maury Island. At its tip was a large industrial site, the Tacoma Smelting Company, covered in piles of black slag. Sander suggested that this was the UFO debris that Crisman had given them, implying, once again, that the whole thing had been an elaborate scam.

Arnold was unconvinced, but he and Smith knew that they had reached a dead end. What had started as a grand adventure had ended only in death, deception and disappointment. It was time to leave. On the way to the airport they decided to drop in on Dahl at the house Arnold had visited when he first arrived. Arnold drove them to what he was sure was the right location, but the house 'looked completely deserted, there wasn't a stick of furniture inside, just dust, dirt and cobwebs everywhere'.[8] Flying home feeling shaken and confused, Arnold stopped off for a fuel-up in Oregon. On take-off his plane's engine cut out suddenly, forcing him to make an emergency landing that bent his wheels on impact. Arnold inspected the engine and found that the plane's fuel valve had been cut off. It was the final calamity in what had been a distressing and frustrating trip.

Whatever this episode was about, it wasn't just flying saucers. It is crucial not to underestimate how seriously the US authorities took the threat of Soviet infiltration at the time. In 1943 the US Army and the FBI had initiated the Venona project to decrypt Soviet intelligence transmissions; it was so secret that not even presidents Roosevelt or Truman knew of its existence. Venona had made its first dramatic breakthrough in December 1946 and the situation it unravelled was nothing short of devastating. To their horror, Venona's orchestrators identified Soviet moles inside the Manhattan Project that built the atom bomb, and several government bodies, including the Office of Strategic Services (which became the CIA in 1947), the Army Air Force, the War Production Board,[9] the Treasury, the State Department, even among President Truman's trusted White House administrators. The United States was paranoid, and with good reason: there really were Reds under the bed, including the four-posters at the White House. As a result, on 21 March 1947, President Harry Truman signed Executive Order 9835, known as the Loyalty Order, which gave the FBI sweeping powers of investigation of all current and prospective federal employees. The resulting Red Scare led to the rise of Senator Joseph McCarthy and the national panic attack of the House Unamerican Activities Committee.

This was America when the UFOs came. So, when their first spokesman, Kenneth Arnold, began talking in the national press about American secret aircraft and atomic power sources, it should have been no surprise that he found himself up to his neck in what looks like a major counter-espionage investigation involving Army Air Force Intelligence – the forerunners of Richard Doty's AFOSI – and the FBI, who filed a lengthy report on the Maury Island case. So was the entire Maury Island incident a set-up to test Arnold? Was Dahl in on the act or just an unwitting stooge? Was the flying 'doughnut' sighting staged for Arnold's benefit? He described the craft as 'balloons' more than once – so perhaps that's really what he saw. Arnold noted that Dahl came across as something of an innocent next to the silver-tongued Crisman, while Dahl referred to Crisman as his 'superior', even though the harbour patrol boat was registered in Dahl's name.

Fred Lee Crisman is central to the mystery. During the Second World War, Crisman had flown with the Army Air Force in south-east Asia and the Pacific and was rumoured to have worked for the Office of Strategic Services (OSS) – a US intelligence agency. After the war, he was an investigator for the Veterans Rehabilitation Association and, in late August 1947, days after the Maury Island fiasco, he applied for a job at the powerful Atomic Energy Commission – guardian of America's nuclear secrets, including its weapons. Maury Island is not far from the Hanford nuclear material processing site, the world's first plutonium site that had provided materials for the Manhattan Project. It has been speculated that the initial 'incident' was actually a cover story for the illegal dumping of nuclear waste in Puget Sound; or was Arnold brought to Hanford to see if he would take an interest in the plant – an interest that would have been shared by his Soviet handlers, if he had them?

Or perhaps we're mystery mongering – as Fred Crisman's FBI file states, Dahl and Crisman simply initiated a scam to extract money out of Ray Palmer for 'flying saucer wreckage', a scam that got them out of their depth when Arnold, himself under investigation, appeared on the scene, dragging the FBI and Army Intelligence along with him. It's possible, but Crisman's involvement suggests that something more complex may have been going on.

Crisman led a short, colourful and eventful life. His name crops up in District Attorney Jim Garrison's controversial investigation into the JFK assassination.[10] Garrison claimed Crisman was working as an undercover agent in the arms world, and may even have been present at Kennedy's assassination. He would later involve himself in Washington and Oregon state politics, where he earned a reputation as a troublemaker and was constantly dogged by rumours of intelligence connections: Crisman certainly fits the slippery profile of a CIA hired hand. He would return to the UFO scene in the late 1960s, a time of increased national interest in the subject, promoting Maury Island as a genuine UFO incident, before dying in 1975, aged fifty-six, leaving a legacy of mystery behind him.

CRASH

The Maury Island affair reveals a keen Intelligence interest in the UFO subject almost from day one. It is now largely forgotten, but another event that took place within days of Arnold's first sighting would become a foundation stone for the UFO myth. The alleged crash of a flying disc at Roswell, New Mexico, in early July of 1947 is the creation story of the American UFO lore, the moment when the scales fell from the American government's eyes and the reality of our place in the universe became clear. According to the myth, the capture of an extraterrestrial vehicle and its occupants was an event of such calamitous consequences that a deep-level cover-up was established, one that remains in place six decades later.

The reality is that the Roswell story as we know it began to take shape only in the late 1970s, and was cemented into place by the publication, in 1980, of *The Roswell Incident* by William Moore and Charles Berlitz. Until then the incident existed only as two or three days' worth of news headlines generated by a press release from Roswell Army Air Force Base, transmitted on 7 July 1947. The original release is now lost, but is thought to have begun as follows:

> The many rumors regarding the flying disc became a reality yesterday when the intelligence office of the 509th Bomb Group of the Eight Air Force, Roswell Army Air Field, was fortunate enough to gain possession of a disc through the co-operation of one of the local ranchers and the Sheriff's Office of Chaves county. The flying object landed on a ranch near Roswell sometime last week.

The story spread quickly around the world, but by the evening of 8 July – the day that most papers ran the story – officers at Fort Worth Army Air Force Base in Texas had identified the remains as those of a 'weather balloon' and its radar-reflecting kite, which were promptly photographed in the hands of base Intelligence chief Major Jesse Marcel. The 'flying disc' story was retracted, the world's press having no reason to suspect that this was anything but the truth, and that was the end of that for at least thirty years.

If the Army Air Force's behaviour in issuing the 'flying disc' press release and then retracting it seems strange to us in hindsight, bear in mind that it did succeed in closing the lid on the story for three decades. This may have been all that was intended, especially if what came down was an item of sensitive, though human, military equipment, even if it wasn't quite the ordinary weather balloon that the Army Air Force's initial retraction would have us believe. Perhaps the most damning pieces of evidence against anything extraterrestrial taking place at Roswell are, ironically, two formerly classified internal documents, one from the FBI, the other from the newly formed US Air Force. The FBI memo, dated 8 July 1947, describes the wreckage being transferred from Fort Worth to Wright Field (now Wright Patterson AFB). The critical section reads:

> The disc is hexagonal in shape and was suspended from a balloon by cable, which balloon was approximately twenty feet in diameter. Major Curtain further advized that the object found resembles a high altitude weather balloon with a radar reflector, but that telephonic conversation between their office and Wright Field had not borne out this belief.[11]

The next item is an internal US Air Force memo sent on 23 September 1947 by General Nathan Twining at Air Materiel Command, who was responsible for Air Force weapons and technology. Written before the Air Force investigations into the UFO sightings had got under way, it is the first official Air Force statement on the subject and demonstrates that at the highest level of control over America's skies, nobody knew what the hell was going on. After stating that the 'phenomenon is something real and not visionary or fictitious' the memo ends with three points for consideration:

> (1) The possibility that these objects are of domestic origin – the product of some high security project not known to . . . this Command.
> (2) *The lack of physical evidence in the shape of crash recovered exhibits which would undeniably prove the existence of these subjects.*
> (3) The possibility that some foreign nation has a form of propulsion possibly nuclear, which is outside of our domestic knowledge. [author's italics][12]

What this internal report, written at the very highest level of the Air Force's command structure and kept secret for many years, doesn't say, is that an alien spacecraft had been recovered at Roswell. So if it wasn't a spacecraft, what was it?

The official US Air Force version of events, presented in *The Roswell Report: Fact Versus Fiction in the New Mexico Desert (*1995), suggests that the Roswell object was a balloon train carrying radar equipment as part of Project Mogul, a highly sensitive mission to gather acoustic data from Soviet atomic bomb tests. It's not an unreasonable suggestion – a Mogul flight was launched from nearby Alamogordo on 4 June and was lost a few days later, while the debris would match rancher Mac Brazel's description of a 'large area of bright wreckage made up of rubber strips, tinfoil, a rather tough paper and sticks'.

The 1995 report was written by Colonel Richard Weaver, who had recently retired from the Air Force. In an unfortunate irony, Weaver's job had been as Deputy for Security and Investigative Programs for the USAF. This meant that he was a disinformation specialist and, in the early 1980s, he just happens to have been one of Richard Doty's superiors at the Office of Special Investigations. This was exactly when Doty and AFOSI were trying to convince Paul Bennewitz that an alien spacecraft *had* come down at Roswell. And the government wonder why the UFO lobby don't believe them!

The General Accounting Office that supervized the 1995 report admitted that their investigation was made difficult because many papers relating to the incident were missing, a fact that served only to encourage those who believed that an ET craft had crashed at Roswell. Might the papers have been destroyed to hide an awkward truth that had nothing to do with aliens? Given the proximity of Roswell to the rocketry range at White Sands, which was shared by all three armed forces, it's likely that whatever did come down probably flew from the test range's direction. There are multiple possibilities as to what it could have been if it wasn't an alien spacecraft or a Mogul balloon but, given some of the terrible things that the US government *has* admitted to doing in the past and subsequently apologized for – like lethal radiation experiments on humans that continued into the 1950s, allowing black men to die of syphilis in the name of science as recently as 1972, or Operation Northwoods, the proposed 'false-flag' plan to bomb its own people in 1962 and blame Cuba – we have to wonder what could have happened at Roswell that was so terrible that we still can't be told about it.[13]

If it was an unconventional balloon or a rocket that crashed, why did Roswell Army Air Force Base transmit the press release that launched a thousand UFOs? Because a saucer crash was considered an innocuous cover that would effectively mask sensitive experiments? We can be sure that the press release was transmitted with specific intent. The 509th Bomber Group stationed at Roswell was perhaps the most elite flying group in the US, if not the world, having dropped the two atomic bombs that ended the Second World War. Being the only atomic bomb squadron in the world, security around its base would have been extraordinarily tight, and slip-ups would have been rare and dealt with very seriously.

Why would such an elite unit, for which tight secrecy was an everyday reality, put out a press release about something as potentially sensitive as a flying disc or even a secret weather balloon project? Why would they mention the incident at all rather than just thank rancher Mac Brazel and ask him to keep his mouth shut as a matter of national security? And if it was an accident, why did base commander Colonel William H. Blanchard, on whose watch the incident took place, end up enjoying a highly illustrious career?[14]

Given the political climate of the time and the press excitement about flying saucers in the weeks following the Arnold sighting, is it possible that the story was deliberately planted? Within the American military there were serious concerns that the flying saucers represented an advanced Soviet technology. Perhaps announcing that one had been captured might send ripples back to the Soviets, ripples that could then be traced by the relevant intelligence bodies. Or perhaps the announcement was intended to lure Soviet moles to Roswell or Wright Field to find out what was really going on? A potentially risky strategy, but one that does make some sense.

Another mystery, and another tantalizing possibility, lies in the pages of an obscure British spy thriller from 1948.

THE FLYING SAUCER

Between about 1930 and his death in 1968, British author Bernard Newman wrote more than a hundred books of fiction and non-fiction, most of them dealing with espionage and war. During the First World War, Newman worked in intelligence and he spent the inter-war years travelling and lecturing throughout Europe, some suspect as an agent, or at least as an informant, for the British government. Although most of Newman's spy novels were potboilers, he garnered the respect of the intelligence world and some of his books were highly regarded for their insight into the games that spies play.

The Flying Saucer, published in 1948, isn't one of them, but it does have the singular distinction of being the world's first UFO book. Its plot concerns a plan by international scientists to stage a fake invasion of hostile extraterrestrials, with the ultimate aim of making the world 'universe-conscious' and bringing about global unity. The book opens with mention of a real speech made by British Foreign Secretary Anthony Eden at a United Nations conference on 1 March 1947. Referring to the encroaching possibility of man-made armageddon, Eden said: 'Sometimes I think the people of this distracted planet will never really get together until they find someone in [sic] Mars to get mad against.'[15]

In Newman's novel the scientists stage a series of 'space missile' crashes at key locations around the world. The second falls in New Mexico and contains mysterious hieroglyphics, a feature that would be attributed to the Roswell debris in the late 1970s. The message is deciphered as a threat by a Martian civilization, and is followed by a missile attack on an uninhabited forest. A later rocket contains the remains of a dead alien – a chimera made up of animal parts – and, in the novel's climax, live 'aliens' land at sites all over the world. As a result of the threatened invasion the USA and the USSR agree to end their enmity and a proliferation of nuclear weapons among even the smallest nations brings peace to the world.

The Flying Saucer seems deliberately to merge fiction and fact; Newman himself appears as a central protagonist, perhaps to hint that it's more than just a story and the book contains a number of other interesting details beyond the central idea of the staged crashes. Newman accurately describes the spread of flying saucer sightings after the first Martian rockets had landed, noting inconsistencies in the variety of shapes, sizes, colours and sounds described by witnesses, something that would frustrate US Air Force analysts in the years to come. He describes a multi-stage rocket design, something that the US Air Force and Navy had both been considering in utmost secrecy; and live human prisoners are exposed to the explosion of an experimental weapon, as was rumoured to have occurred during America's atomic tests. As part of the aliens' arsenal, the book's key scientist, Drummond, develops a portable electromagnetic ray that can stall aircraft and car engines:

> Drummond's apparatus here represented no new idea, but a development of an old one. For years scientists have known how to stop an engine, by interference with its electrical processes, by means

> of the emisssion of a charge or ray. The difficulty was that a huge and complicated apparatus was necessary, and that its range was short: thus much simpler methods of stopping an engine were available – putting a bullet through it, for example.

Car-stopping rays and destructive 'death rays' were part of the technological folklore of the Second World War; a death ray was implicated in the death of Kentucky National Guard pilot Thomas Mantell, who died on 7 January 1948 pursuing a UFO that turned out to be a secret Navy Skyhook aluminium balloon. Car engine failures and radio interference also became a key element in UFO accounts in the mid-1950s: could this have been related to Drummond's ray?

Did Newman know, or think he knew, something about the Roswell Incident? Had he gleaned inside information from his contacts in the intelligence world? Who knows. Yet Newman certainly understood how folklore worked. When a pilot in the book fakes a story about a UFO crash, a new wave of reports is triggered, even after his public confession, leading Newman to remark: 'Once a story has been started, it is never entirely quashed.'

FOUR
LIFT-OFF

'My task was to prepare the mind of the people well in advance – to make it receptive to ideas about other planets. To this end I stimulated articles in the popular press all over the world. I revived old controversies about canals on Mars and even the unexplained white streaks on the Moon'

Bernard Newman, The Flying Saucer *(1948)*

A couple of months after John and I began thinking about our film I received an advance copy of a book called *Project Beta* by Greg Bishop, a writer and researcher from Los Angeles. Greg and I had met a few times and instantly hit it off, sharing a keen interest in the overlaps between ufology, mysticism and pop culture. In researching the book, which described the US Air Force campaign against Paul Bennewitz, Greg had met Doty for several hours at a Denny's restaurant in rural New Mexico. Greg hadn't been allowed to record their conversation or to take notes, but it was still a breakthrough – the first time that the story had been heard from the Air Force's perspective.

Project Beta also gave us one critical piece of data that, as far as we knew, nobody had ever had before: a clear photograph of Doty, pictured in black and white in 2000, apparently in a photographic studio and looking, well, astoundingly normal. Here was a man in his late forties in jacket, tie and striped shirt, a lozenge-shaped face on gently sloping shoulders topped by a short crop of policeman's hair. The smile was awkward, made quizzical by a slight upturn to the left of the mouth, echoed by a faintly raised eyebrow above. A film of sweat over his face adds to the overall sense of anxiety. But perhaps I was reading too much into this first image of our quarry: it may have been a hot day in New Mexico and, let's be honest, nobody likes having their photo taken.

On the back of *Project Beta*, in early 2005, Greg Bishop appeared as a guest on America's fearsomely popular *Coast to Coast* late-night talk radio show. *Coast to Coast* both reflects and dictates what forms of weirdness are currently at large in the American imagination, particularly the imagination of those Americans who are up in the small hours listening to talk radio. UFOs are a perennial *C2C* fascination, with several hours a week dedicated to the subject, followed by the usual ghosts and hauntings, bigfoots, psychic powers and healing, gateways to hell and the ever-looming apocalypse, whether it be via UN invasion, fuel shortages, an ET takeover, the return of the Mayan gods or a collision with Planet X. It's an unending panorama of strangeness and suspicion.

On the show Greg spoke about his book and the role that disinformation has played and continues to play in the UFO arena. He then introduced his special guest, none other than Richard C. Doty, calling in live from New Mexico. Now we had a voice to add to the photograph; unexpectedly high-pitched, geeky and prone to outbursts of snickers and giggles. Doty introduced himself as a private citizen and insisted that his activities in the UFO field had ended in the 1980s. He spoke of his friendship and respect for Paul Bennewitz, his sadness at what had happened to him and how he would like to have been able to prevent his nervous breakdown. And then he dropped the bombshell. After everything he'd seen and done as a seeding agent for bizarre beliefs, Richard Doty himself believed that there was an extraterrestrial presence on Earth.

Here was a man whose name was almost synonymous with deception about aliens and UFOs, and he was trying to tell us that it was all true. And not only that, but he expected us to believe him. Did Doty have such a low opinion of *Coast to Coast*'s listeners that he thought he could get away with such a stupendous *volte face*? Was he having a private joke with himself at the world's expense? Was he still under oath from his AFOSI days? Was he still active as an intelligence agent and going on the air was all part of a day's work? Or perhaps, most outrageously of all, he really did believe it; perhaps he really did know something or had seen something that had changed his mind. Or maybe over the years the UFO material had started to get to him: liars sometimes begin to believe their own lies; sometimes they *become* their own lies. Or perhaps this had been the case all along, and that was why US Air Force Intelligence had hired Doty, recognizing in him someone who could merge so completely with a role that he would begin to lose his sense of self, like a method actor. For now, anyway, we couldn't believe our luck. We had an image and a voice of Richard Doty; or, at least, someone who said that they were Richard Doty.

Following Doty's radio appearance I kept an eye and an ear out for signs of the UFO's return from oblivion, but it wasn't happening. While the Internet was overrun with UFO websites, they all told the same old stories and reran the same old fuzzy photographs that had been discredited years ago. The subject needed a jolt of electricity, a new plot twist to bring it lumbering back to life. I feared that it had been left in suspended animation for too long and wondered whether UFO enthusiasts would soon seem as quaintly anachronistic as ham radio buffs or train spotters. Then a miracle happened. It started small, as a ripple in an obscure Internet backwater, but it soon became a wave. Something was stirring in UFOland.

REQUEST ANONYMOUS

'First let me introduce myself. My name is Request Anonymous. I am a retired employee of the U.S. Government. I won't go into any great details about my past, but I was involved in a special program . . .'

So began the email received on 1 November 2005 by Victor Martinez, a substitute teacher on America's west coast who ran what must have been one of the most remarkable email groups on the Internet. Its 200 or so members made up a veritable *Who's Who* of scientists, military personnel and Intelligence officers known to have had an interest, or more often than not, a direct involvement, in the UFO phenomenon over the past thirty years. Here were one-time and current employees of the CIA, Defense Intelligence Agency (DIA) and the National Security Agency (NSA), US government remote viewers, free energy researchers, theoretical physicists and venture capitalists, plus a healthy smattering of mystics, witches, alien contactees and abductees.

Then there was Request Anonymous – known simply as Anonymous. According to Martinez, Anonymous had been monitoring the list for about six months before introducing himself and his 'special program'. So who was he? Allegations immediately began to ping back and forth, but Anonymous managed to keep his identity a secret. Martinez didn't attempt to unmask him. 'If word ever got back to [Anonymous] that I was trying to ID him, he would simply have packed up his bags and found another UFO list moderator to release his incredible story through.'

Incredible was the right word for Anonymous's voluminous account, which grew by thousands of words with each passing month. When the emails stopped coming, three years later, they encompassed thirty-one timed 'releases' and several tens of thousands of words. And these were just extracts from the alleged original document which, according to Anonymous, was a highly secret 3,000-page report compiled by the DIA in the late 1970s. Where such a hefty tome could be found, and how Anonymous got hold of his copy, remains a mystery, but it wasn't at his local library.

This, in drastically reduced form, is what the releases had to say:

In 1947 two ET spacecraft crashed in New Mexico – this is the event known since the late 1970s as the Roswell Incident. Six ETs died in the crash and one survived. The remains of their craft were taken to Wright Patterson AFB in Dayton, Ohio, while the surviving ET, nicknamed EBE 1, was installed at Los Alamos Laboratories in New Mexico, where he lived until 1952. During this time he attempted to make contact with his home planet. Sadly for EBE 1 they returned his call only after his death, but it was a historic moment for America. From this point on, the United States government was in regular communication with an extraterrestrial race, the Ebens. The only problem was that they couldn't tell the rest of the world about it.

In late 1962 President Kennedy – who, some say, was killed before he could reveal the truth about UFOs to the American public – authorized a foreign exchange of cosmic proportions. A team of twelve specially trained humans whose identities were subsequently erased (or 'sheep-dipped' as they say in the intelligence business), would return with the Ebens to their planet in a programme called Project Crystal Knight. Preparations were made for a face-to-face meeting between Eben and human ambassadors and on 24 April 1964 two Eben spacecraft entered Earth's atmosphere. One of them landed close to Holloman Air Force Base in New Mexico. A team of senior US government officials boarded the craft and were presented with a holographic device known as the Yellow Book, which contained a complete history of planet Earth. The personnel exchange was agreed for the following year and in July 1965, the human Away Team entered an Eben craft while another ET, nicknamed EBE 2, stayed behind.

The ETs' planet, known as Serpo by the human visitors, is thirty-eight light years from Earth, in the Zeta Reticuli star system.[1] Serpo, according to Anonymous's own report, is a little smaller than Earth and has two suns. It is hot, flat and dry; harsh but habitable, especially in the cooler northern regions where the humans made their home. The Away Team spent thirteen years on Serpo where, despite a few misadventures, they were welcomed by the Ebens and given freedom to roam. There were approximately 650,000 Ebens on Serpo, living in about a hundred small, autonomous communities around the planet. There was no centralized government, though there was one large, central community that served as a hub for Eben industry and resources. Everybody worked and, in return, was supplied with everything they needed in order to live spartan but happy lives. Crime was non-existent in this quasi-socialist utopia, but war was not. Three thousand years ago the Ebens had fought a vast, hundred-year interplanetary conflict with the civilization of another planet, which they annihilated at the cost of making their original home planet uninhabitable. Since then, the Ebens had been intergalactic drifters, visiting a number of other species and civilizations, including our own, before settling on their current home planet.

By the time the human team returned to Earth in 1978, two of them had died. Two chose to remain on Serpo and stayed in contact with Earth until 1988. Those team members who did come home had been exposed to high doses of radiation thanks to Serpo's twin suns, and it was this that ultimately killed them – the last dying in Florida in 2002.

Curiously, while Anonymous provides his audience with a wealth of detail about the Ebens' culture, their living habits, even their digestive systems, we are never given a description of what the aliens actually look like. Defenders of Anonymous's tale suggested that this was because the original people for whom the report was intended – those on the inside of the US UFO agenda – already knew what the ETs looked like; there was an Eben in captivity at Los Alamos, after all. The full report was also said to include a number of photographs, which Anonymous promised to share with the world, including one of the Ebens playing a football-like game. This particular image never appeared, though after a few months some desert landscapes featuring twin suns did emerge, only to be swiftly dismissed as Photoshop creations, and poor ones at that.

Rather than being met with incredulous laughter, Anonymous's initial message was verified by two members of Victor Martinez's email list, Paul McGovern and Gene Lakes (sometimes called Loscowski), both of whom provided further background material to support Anonymous's claims. McGovern was identified as a former DIA security chief, stationed at Area 51; Lakes appeared to be another DIA insider. But there was a problem: while McGovern and Lakes's credentials sounded impressive, their identities couldn't be verified outside the confines of Martinez's email list. This shouldn't have been too much of a surprise if, as the two men claimed, they had spent their careers working in the black world of military intelligence. A bigger problem was the other person backing up both the Serpo story, and the identities of Paul McGovern and Gene Lakes: Richard C. Doty.

Over the next few weeks Serpo began causing a splash on the Internet. It even garnered a mention in a London commuter newspaper. With its terse, militaristic, first-person narrative, the Serpo papers combined the intensity of a Tom Clancy thriller with the pulp charm of an Edgar Rice Burroughs space opera. They occupied that curious space, neither entirely fact yet not quite fiction, shared by all otherworld narratives, whether they be tales of terrestrial explorers such as St Brendan and Marco Polo or the heavenly travels of so many holy men. Each new release worked like an episode in an old RKO serial, *King of the Serpomen* perhaps, bringing with it new promises and new cliffhangers: we would see photographs; we were going to get the names of the exchange crew; we would hear from a living survivor. We never did, of course.

If it served no other purpose, the Serpo affair, after years in the doldrums, had finally caught the attention of the UFO community, fanning the dying embers of enthusiasm back into roaring life. On *Coast to Coast*, the famed alien abductee and horror fiction author Whitley Strieber recalled meeting an old military man at a UFO convention in the 1990s. The man told Strieber that he had been to another planet before muttering what Strieber thought was the word 'Serpico'. (*Serpico* is the name of a 1973 police thriller starring Al Pacino). Now, a decade or so later, it all made sense to Streiber, and – with Strieber's support, and the continuing flow of information and chatter over the airwaves and on the web – Serpo was starting to feel as if it might actually have happened. More miraculously, for the first time in several years, UFOs were once again a hot topic.

As the Serpo excitement spread, an English-Canadian personal development trainer named Bill Ryan offered to set up a website as a clearing house for Anonymous's information. Bill had stumbled upon the Serpo material quite accidentally, after his girlfriend forwarded him emails from an anti-George W. Bush mailing list that also happened to be run by Victor Martinez. Initially Martinez had kept his political and UFO mailing lists separate, but the Serpo material was so momentous that he felt he had to share it with the world, pinging it out to his political list and so drawing in Bill.

Bill was no stranger to fringe beliefs. Although he didn't initially admit it, he was once a keen Scientologist. So keen, in fact, that he had joined Ron's Org, a splinter faction of the retro-science-fiction religion who felt that founder L. Ron Hubbard's original teachings were being ignored by the new generation of Scientologists. Bill also admitted to having dated a woman he believed to be an extraterrestrial. Despite this, and the extraterrestrial origins of L. Ron's religion, Bill was a newcomer to the UFO scene. Bill quickly became a passionate advocate of Anonymous's story and devoted all his time and energy to maintaining the website, which soon attracted huge numbers of visitors. Whether he intended it or not, Bill Ryan was now Serpo's front man – and John and I needed to talk to him.

Arranging to meet Bill was straightforward; flushed with the evangelistic spirit, he was only too happy to spread the word about Serpo, and came to meet John and me in London in December 2005. Awaiting his arrival, I realized I wasn't quite sure what to expect from Bill. Whatever I expected, it was not what we got: Bill was not your ordinary management consultant; he was an effusive late-fortysomething with friendly, weathered features framed by thinning shoulder-length reddish hair. From the soles of his shoes to the top of his battered felt outback-hat, his clothes were worn out and full of holes.

For someone who claimed to have worked for the likes of Hewlett-Packard and PricewaterhouseCoopers, Bill certainly cut a casual image. Now I'm not usually one to pick holes in somebody else's appearance – my own garb on any given day is full of them – but Bill looked as if he'd been living out of a car for several days, probably because he had. Working so intensely on the Serpo story had put a serious strain on Bill's relationship with his girlfriend, and for the time being he wasn't sure where was home (need I remind you that UFOs can seriously damage your lifestyle . . .).

Bill was absolutely sincere in his convictions about Serpo; the story had consumed him entirely. While being interviewed he was constantly checking his Apple laptop (battered, naturally) for emails. Rumour had it that a new Anonymous release was due any day, and Bill was feverish with anticipation for the next instalment, while also replying diligently to the steady stream of Serpo-related questions that were now coming his way.

Bill's intentions may have been genuine, but he was new to the UFO business, and it wasn't going quite as smoothly as he may have anticipated. Astronomers had raised questions about the orbital data contained in the report (allegedly provided by celebrity astronomer Carl Sagan) but there were other, more basic issues nagging at me. My main concern was that, beyond the main event of the ET–human exchange, there really wasn't anything in the Serpo story that wasn't already present in the UFO lore. Any movie buff could also point out the obvious parallels to Steven Spielberg's UFO epic *Close Encounters of the Third Kind*, which climaxes with the landing of a colossal disco-ball UFO at a secret site near Devil's Tower in Wyoming. Here Richard Dreyfus's character joins twelve military personnel who board the ET craft, presumably to be taken to the benevolent aliens' planet. Could it have been Serpo? Many in the UFO community believe that both *Close Encounters* and Spielberg's *ET* were made to acclimatize us to the truth about alien visitation, a process begun in 1951 with *The Day the Earth Stood Still*. If this was the case, we have to wonder what Spielberg was trying to tell us with his apocalyptic take on *War of the Worlds*, but to Serpo-watchers the release of *Close Encounters* in 1977, just a year before the Serponauts' return, was no coincidence.

Rather than sensing a snake in the grass, Bill felt that Serpo's inconsistencies actually detracted from the likelihood of it being a hoax, reasoning that someone who would take the time to fashion such an elaborate tale would have made the effort to get their astronomical facts straight. I wasn't so sure – what if they didn't know any astronomical facts in the first place?

'I'm not trying to browbeat anyone into believing this,' explained Bill, 'only to consider it as a possibility, but I think a simple hoax or a prank can absolutely be ruled out. It's too complex for that, and there's too much circumstantial corroboration. Misinformation falls into the same category – that would mean it's all false. But it could be disinformation. That means part truth, part fiction. And the fiction part could be as little as 5 per cent for the entire story to be thrown off-kilter.'

Disinformation. Noise. Was this what Serpo was about? An attempt to use the Internet to seed information within the UFO community, much as AFOSI had done, using faked documents in the Bennewitz affair? Were we seeing an obscure new salvo in the information war? Were Serpo and the Martinez list a vivarium for information? Was Serpo a sociological or psychological research project being carried out by one or other intelligence agency, or a university – perhaps by a member of Martinez's list? We might think of it as memetic tracking: following the paths that information follows over the web would be a useful exercise in our data-saturated age. Like attaching a transmitter to a whale, or tracking a barium meal through a hospital patient's digestive system, it can teach us a lot about both what is being followed and the territory it's travelling through – in intelligence lingo, this is called a 'marked card'.

If Serpo did have its origins in the intelligence world, and it looked to many observers as if it might, it may have had nothing to do with UFOs. Perhaps sensitive information was encoded within the releases, or perhaps it was what the intelligence world calls a 'false flag' operation, made to look as if it was the work of UFO insiders and intended to lure foreign or industrial spies into its web. Was it just a coincidence that there were so many intelligence and military personnel on Martinez's UFO list? Was Serpo an attempt to flush somebody out of hiding?

One interesting idea raised online, then swiftly shot down, was that Anonymous had actually stumbled on to genuine government documents, but ones that had originally been created in order to fool somebody else – the Russians, for example. One could see the Ebens's blissful-yet-spartan communal existence appealing to the Russian political machine of the 1960s or 1970s as a cosmic, utopian vision of their own world. Another possibility was that the Serpo material was the work of Alice Bradley Sheldon. Sheldon worked for US Air Force Intelligence in the 1940s and as a CIA operative in the 1950s before becoming celebrated as a new-wave science-fiction writer under the pseudonym James Tiptree Jr, a secret she kept until 1977. Had Sheldon, with her talent for writing SF and her CIA connections, been hired to write the Serpo documents in the 1960s or 1970s? Were they part of a convoluted disinformation game to throw the Russians off the scent of another project, like the US Air Force's Manned Orbiting Laboratory (MOL), a highly secret spacebound orbital reconnaissance programme begun in 1963?

There are certainly parallels between this extremely advanced (for the time) project and the Serpo story. Seventeen US Air Force men were prepped and primed for life, a month at a time, in the space vessel's cramped confines, and no one but they and their supervisors knew what it was they were training for. The programme was cancelled after a single test flight – the costs involved were astronomical and the level of potential danger for the crew ended up being unacceptably high, especially when unmanned satellites could now do much of what the the human crew would be doing: Sheldon could have written the documents and even the MOL timeline. Even the MOL timeline almost fits, with the programme being initiated in 1963 and projected as running into the mid-1970s, parallel to both the short life of James Tiptree Jr and the alleged Serpo exchange.

Sadly none of it is true. The Sheldon/Tiptree suggestion was made by an anonymous emailer in the early days of the Serpo saga, while the MOL embellishment is my own. The persona behind the Sheldon titbit later announced that they had invented the whole Serpo saga as part of a university sociology course – itself an infuriatingly plausible possibility – before vanishing for ever into the digital void.

Whatever its origins, Serpo provided the UFO culture with a much-needed shot in the arm and brought many of its key players from the previous decade's heyday crawling out of the woodwork. It also gave rise to a pair of extremely popular UFO and conspiracy theory online message boards, *Open Minds* and *Above Top Secret*. These acted simultaneously as receivers and transmitters for UFO information and, like Martinez's list, corralled disparate UFO hunters and parapolitical enthusiasts into one place. Here people shared views on everything from the origins of the 9/11 attacks and secret US government bases on the Moon, to the latest developments in military technology. A useful place then for both American and international intelligence operators to hang out and monitor geeks and extremists of many stripes. And, perhaps, to get involved.

A news story from late 2008 shed some light on the way that intelligence agencies were using message boards and forums to conduct their operations.[2] Some time in 2006 a hacker using the handle Master Splynter joined a web forum called Darkmarket, a leading clearing house for credit-card hackers and data thieves. Here people could buy and sell data and the technology required to reap data and produce fake credit cards. Over the next few months Master Splynter gradually took over operations until, in October 2008, he declared that he was shutting down Darkmarket:

> It is apparent that this forum ... is attracting too much attention from a lot of the world services (agents of FBI, SS, and Interpol). I guess it was only time before this would happen. It is very unfortunate . . . because . . . we have established DM as the premier English speaking forum for conducting business. Such is life. When you are on top, people try to bring you down.

Master Splynter knew this because he was actually FBI cybercrime agent J. Keith Mularski, who had infiltrated the site to shut down a large international credit-card fraud ring. Whether *Above Top Secret*, *Open Minds*, or indeed any of the myriad other UFO and conspiracy sites owe their existence to similar operations we can't say, but wherever advanced military hardware and UFOs are being discussed, the intelligence world will always be listening.

Beyond launching websites, Serpo also turned Bill Ryan into an instant ufological celebrity. Within a few weeks of setting up his site, Bill was invited to be a keynote speaker at the 2006 Laughlin International UFO Convention, one of the biggest such gatherings in the USA, if not the world. The Laughlin Convention would take place in March and, as if to flag up Bill's presence, the February 2006 issue of America's *UFO Magazine*, the only news-stand UFO publication in the English-speaking world, ran a special issue about Serpo. Complementing a piece by Bill was another article. It began:

> My name is Richard Doty retired special agent, Air Force Office of Special Investigations (AFOSI), and now a private citizen living in New Mexico. I've been an avid reader of *UFO Magazine* for the past several years . . .

> In early 1979, after arriving at Kirtland Air Force Base as a young special agent with AFOSI, I was assigned to the counter-intelligence division of AFOSI District 17. I was briefed into a special compartmented program. This program dealt with United States government involvement with extraterrestrial biological entities. During my initial briefing I was given the complete background of our government's involvement with EBEs. This background included information on the Roswell incident . . . basically, this was exactly the same information that Mr Anonymous released.

For a former intelligence agent and a man who had effectively vanished for more than a decade, Doty had become quite cavalier about being in the public eye. He was also communicating regularly on Victor Martinez's email list, chipping in with pieces of information that backed up Anonymous's claims about Serpo.

John and I decided that it was time to make contact. In our email to Doty we explained that we were making a film about the involvement of the intelligence agencies in the UFO subject, and would like to talk to him about his experiences. We mentioned that we would be going to the Laughlin UFO Convention to film Bill Ryan. Perhaps we could meet Doty in New Mexico?

A reply was not long in coming. Doty would consider an interview with us, perhaps in Albuquerque, but better than that, he was going to be at Laughlin and would be happy to meet us there.

There was only one thing for it: we were going to Nevada.

FIVE

CONFERENCE OF THE BIRDS

'What power urges them at such terrible speeds through the sky? Who, or what, is aboard? Where do they come from? Why are they here? What are the intentions of the beings who control them? . . . Somewhere in the dark skies there may be those who know'

'Have We Visitors from Space?' LIFE *magazine, 7 April 1952*

A single two-mile strip of hotels and casinos, Laughlin, Nevada tips out on to the Colorado River that separates the southernmost tip of Nevada from Arizona on the east and California on the west. Founded by a former Minnesota fur trapper, Don Laughlin, in 1964, the town is now the third most popular gambling destination in Nevada and attracts what are diplomatically known as the 'low-rollers', those weekend gamblers who don't want to splash their cash on nude magic shows and Bette Midler concerts, either because they're locals, or because they can't afford to. Laughlin is all about slots and tables, and really there's not a whole lot else to do there, though it attracts plenty of visitors.

Every April since 1983 it has hosted the Laughlin River Run, a gathering of as many as 70,000 motorcycle enthusiasts and gangs. Unfortunately it's this that has earned the town its place in history, as the site of the 2002 River Run Riot, a pitched battle between the Hell's Angels and Mongols biker gangs that left three dead and many more seriously hurt.

Then there's the International UFO Convention. Held here since 1991, this has become one of the largest such gatherings in the US. The 600 or so ufologists and UFO enthusiasts who make it every year probably don't consume as much beer, gas or amphetamines as the River Run attendees, but what they may lack in grizzle and hard living, they make up for in starry-eyed enthusiasm.

It was February 2006. John, Zillah Bowes (our camera operator), Emma Meaden (our sound recordist) – both recent graduates of John's film school – and I were driving the ninety miles down from Las Vegas to Laughlin. Our aims were to follow Bill Ryan's progress as Serpo's figurehead, to talk to Greg Bishop, who was speaking at the convention about the Paul Bennewitz affair and, ultimately, to track down and interview Richard Doty.

As we rushed across the monotonously empty, red Martian desert, roadside markers became our only reference points in space and time: fierce concrete dinosaurs, looming cacti, hand-painted billboard ads for meteorites, Indian jewellery, petrified tree stumps and scorpion paperweights. Roadside shrines – moments of high-speed tragedy preserved in the dry heat of eternity – littered the hard shoulders, marked by white crosses, plastic flowers and weather-worn cuddly toys, all arranged neatly alongside the ripped metal and shredded rubber remains of their vehicles.

The debris set me thinking again about the UFO crash at Roswell, reverberating out through the decades like amplified soundwaves, eventually becoming many times larger than the original point of impact could ever be. Roswell has become one of the great myths of the American West, a tragedy of more-than-human proportions. A tale of innocence, even paradise, lost and the triumph of fear and secrecy over hope and truth at the birth of modern America; and it is what had brought us, and the other UFO Convention attendees, to this godforsaken place.

Gliding over a rocky bluff we saw the glittering landing-strip lights of Laughlin stretch out before us. Our destination was the Flamingo Hotel, its twin towers colossal and luminous like space stations locked in binary orbit in the black desert night. We parked our car and entered the hotel beneath what appeared to be a forty-foot-tall pink neon vulva. Once inside, the womb was no longer a metaphor but an inescapable reality.

Everything you needed was here, the hotel providing a self-enclosed, air-cooled environment that kept you sealed off from the hostile desert terrain outside. And even if you wanted to leave, the song of the sirens would only call you back. The signature tunes of each slot machine, lined up in their hundreds just a few feet from the main entrance, slid into each other to form a continuous fluid harmony as nourishing to the soul as the Colorado River outside must have been to the pioneers who first explored this region. It is amniotic music, like the sound of a thousand toddler toys; comforting, reassuring, overpoweringly sweet. Everywhere you looked people were pinned to their seats in a constant, rolling state of low-level epileptic seizure, winking lights reflected in glazed, unblinking eyes. Some were connected to their machines by coiled rubber umbilical cords attached to plastic credit cards, all grasped large tubs of coins in one hand, even larger tubs of coffee, beer or fizzy drinks in the other. There was no time in here, no clocks, no sense of day or night, and no geography; once you got past the first row of machines all the exits on the huge gaming floor were obscured by jagged corridors of light. Anyway, why would you ever want to leave?

Today is Sunday, I reminded myself. The UFO Convention was to last a week, and for that week, this would be our home. I waded through the list of topics that we'd be hearing about at the conference: sub-aquatic alien bases off Puerto Rico and California; UFO crashes in Nova Scotia, Roswell and Aztec; the ET origins of humankind; crop circles; the coming apocalypse in 2012; cattle-mutilating invisible aliens and wormholes at a Utah ranch;

ET-human hybridization programmes; UFO propulsion devices. Little had changed in the decade since I had last been immersed in the material.

I realized what connected the UFO Convention to the casino: faith. Faith that the next coin would be the one to make your fortune and change your life; faith that the ETs are here among us, preparing to reveal themselves and change all our lives. And if you didn't have faith when you entered the Flamingo, after a week of the UFO Convention, you would surely believe by the time you left.

The UFO Convention took place in three parts. There was a large room in which the talks were held, a step stage at one end and a few hundred faux red-velvet and gold dining chairs in front. The room was rarely full and it seemed that a good proportion of the audience was nodding off at any given time. The lecture space opened out into the main foyer where the real action took place: micro-conferences around coffee-and-cake-strewn tables, networking, hushed exchanges – 'I was playing billiards with my mind, man'; 'I used to be into beams, but now I'm into orbs, I get them every time I take a picture'; 'I used to be a nuts-and-bolts man, but now I think it's just whatever it takes'.

Room three was the all-important dealers' room, where beliefs were bought, sold and reinforced. The UFO literature and media pool is absolutely vast and the thousands of books, DVDs and even near-extinct VHS tapes on display here represented merely the leading edge of a mothership-sized circulating library. Here were accounts from first-hand UFO experiencers describing what had happened to them, researchers describing what had happened to the experiencers, scientists explaining what UFOs are and why they are here, scientists explaining what UFOs aren't and why they're not here (very few of these, naturally), channellers explaining where the UFOs come from and why they are here, conspiracy theorists revealing where on Earth the UFOs are hidden, aliens explaining what UFOs are, where they come from and why they are here. It's a swarm of data: not information but infestation. It's almost impossible to pick your way through all the material and form a coherent narrative for the UFO story; there's no elegant five-note solution like the finale of *Close Encounters of the Third Kind*, because the signal – if there even is a signal – is lost amid the cacophony.

To clear my head I returned to the foyer and noticed that the smaller meeting rooms around its edge were all named after exotic birds: there's a toucan, a parrot, a bird of paradise. This couldn't be more perfect. In the mid-1970s a loose-knit group of scientists, intelligence specialists and military personnel, all fascinated by the UFO subject, began trying to use their insider knowledge and contacts to make some sense of the situation. Almost all of them were now on Victor Martinez's email list. Over time these insiders had developed a distilled version of what their investigations had uncovered and they called this 'The Core Story'. Much of this core story forms the basis for the Serpo material: the Roswell crash, EBE the ET survivor, the landing at Holloman Air Force Base. When researcher Bill Moore was drawn into this network by Richard Doty and his colleagues at AFOSI, he gave them the nickname the Aviary and identified them all using bird names. And here we were, preparing to meet Richard Doty in Laughlin's own aviary.

The conference was more revival meeting than science fair or intelligence briefing, so what had persuaded these government insiders that UFOs were more than a rumour and were worth taking seriously? Something must have happened in the three decades between Kenneth Arnold and *Close Encounters of the Third Kind*; they must have seen or heard something. But what?

WHERE DO THEY COME FROM?

The first flush of UFO sightings in 1947 had contributed to a growing sense of unease brewing within the nation's military command. Until Pearl Harbor, America had considered itself invulnerable to surprise attack, but things were different now. The ghost rocket mystery and the ongoing spate of flying saucer sightings created real concerns about the country's vulnerability to aerial infiltration, while the longer-term prospect of a Soviet atom bomb just didn't bear thinking about. Such fears were no bad thing for the newly formed Air Force: securing the nation's skies was its job. As long as people remained afraid of an aerial attack then the US Air Force wouldn't have to worry about budget cuts, or losing funding to their arch rivals, the Navy.

In the wake of General Twining's 23 September report – the Air Force's first official statement on UFOs – it was decided that the Air Force would begin a discreet official investigation of the flying saucer problem, called Project Sign, with the aim of finding out what they were and where they came from. Project operated from Wright Patterson Air Force Base, home of the US Air Forces's Air Technical Intelligence Center (ATIC), which specialized in studying and, wherever possible, back-engineering foreign technology. They had plenty of captured German aircraft to play with and a large team of German engineers, rounded up in secret at the close of the Second World War, to make sense of the trunks full of technical papers that came with them.

Project Sign's greatest concern was that the saucers were Soviet aircraft derived from Nazi experimental designs but, after studying their German blueprints for clues and comparing them with reports of the saucers' astounding flight capabilities the project team reached another, more shocking conclusion: they weren't ours. That is, they weren't human.

By autumn 1948 the Sign team had prepared a top-secret *Estimate of the Situation*, published at about the same time as the first issue of Ray Palmer's new magazine, *Fate*, the cover of which featured a spectacular image of flying saucers. Nobody is sure what the *Estimate* actually estimated, but by the time it reached Air Force Chief of Staff Hoyt Vandenberg he was so enraged that he ordered all copies to be destroyed. As a result, Sign's final report, issued in February 1949, played down the extraterrestrial hypothesis (ETH). A supplementary paper by Dr James Lipp of the RAND Corporation think tank summarized the ETH's inadequacies, and his conclusions are as relevant today as they were sixty years ago:

> It is hard to believe that any technically accomplished race would come here, flaunt its ability in mysterious ways and then simply go away . . . The lack of purpose apparent in the various episodes is also puzzling. Only one motive can be assigned; that the spacemen are 'feeling out' our defenses without wanting to be belligerent. If so, they must have been satisfied long ago that we can't catch them. It seems fruitless for them to keep repeating the same experiment . . . Although visits from outer space are believed to be possible, they are believed to be very improbable.

Lipp's final remark is interesting, given that in May 1946 he had been one of the authors of *RAND Report No. SM-11827, Preliminary Design of an Experimental World-Circling Spaceship*, a plan for a multi-stage space rocket based on the German V-2, developed in response to the Navy's own rocket designs.[1] The report opened with the following prediction:

> Though the crystal ball is cloudy, two things seem clear:
>
> 1. A satellite vehicle with appropriate instrumentation can be expected to be one of the most potent scientific tools of the twentieth century.
>
> 2. The achievement of a satellite craft by the United States would inflame the imagination of mankind, and would probably produce repercussions in the world comparable to the explosion of the atomic bomb.'[2]

With Sputnik and Telstar still a decade away, flying saucers were early harbingers of the space age now dawning over America.

In February 1949 Project Sign was secretly transformed into Project Grudge, with the express purpose of publicly playing down the flying saucer sightings and diminishing public enthusiasm for the subject. To this end they assisted Air Force-friendly journalist Sidney Shalett in writing a two-part article for the *Saturday Evening Post*, 'What You Can Believe about Flying Saucers', published on 30 April and 7 May, just days after the US Air Force presented its own top-secret saucer summary to the country's intelligence bodies. Shalett's piece was the US Air Force's first detailed public statement on the flying saucer subject and did a good job of representing Grudge's disaffection for the subject. To the Air Force's relief, their strategy seemed to work: by the end of the year it seemed that they might finally have washed their hands of flying saucers.

But their peace would be short-lived. The December 1949 issue of the popular men's magazine *True* ran a sensational article by former Navy Marine and pulp-fiction author Major Donald Keyhoe. The title said it all – 'Flying Saucers are Real' – while its first line left nothing to the imagination and blew apart any hope that the Grudge team might have had about an early retirement: 'For the past 175 years, the planet Earth has been under systematic close-range examination by living observers from another planet.'

It was a sellout for *True* and launched the career of Keyhoe, the phenomenon's most vociferous proponent. *True*'s editors wasted no time in capitalizing on the issue's success, following it up in March 1950 with a piece penned by another Navy man, Commander Robert B. McLaughlin, who ran the Navy section of the White Sands Missile Testing Range at Alamogordo, New Mexico. 'How Scientists Tracked a Flying Saucer' described a spectacular sighting by Navy scientists monitoring a ballon launch with a theodolite. McLaughlin was 'convinced that it was a flying saucer and, further, that these discs are space ships from another planet, operated by animate, intelligent beings'. McLaughlin then fully socked it to Project Grudge – 'Hallucination? Optical illusion? I think it is reasonable to say that illusions do not appear simultaneously and identically to five different trained weather observers' – before going on to describe a saucer sighting of his own.

These two reports, by respected senior members of the military establishment, immediately undid all the work that the Air Force had put into deflating flying saucers. More damagingly, they made what had, until then, been largely the domain of mystics and Ray Palmer's science-fiction fans a respectable topic of research and conversation for millions of educated, serious-minded Americans.

Yet all was not as clear as first appeared. Both Keyhoe and McLaughlin were Navy men: their wild claims about extraterrestrials intruding into American airspace dramatically undermined the Air Force's attempts to assert their authority over the flying saucer subject and, more crucially, their role as protectors of the skies. Could the timing of these articles have been intended to cause the Air Force problems? In 1948 the Navy had generated a self-serving press scare by announcing that Russian subs had been spotted patrolling the Pacific. Were they now using similar tactics to embarass the Air Force and foment suspicion of its activities, and its leadership?

That there was no love lost between the US Air Force and the Navy after the Second World War was no secret; their enmity had erupted into all-out warfare in early 1949, resulting in a House Armed Services Committee hearing that spring. The feud reflected a deep divide over military strategy: the Air Force insisted that any future war would be won solely through the use of strategic bombing with nuclear weapons. Such a strategy, they argued, rendered the Navy obsolete and many at the Department of Defense agreed with them. When Secretary of Defense James Forrestal, a Navy man, quit his post in April 1949, his replacement, Louis A. Johnson, an Air Force supporter, cancelled construction of a proposed Navy 'supercarrier' in favour of the Air Force's huge B-36 bomber. The Navy retaliated by leaking documents accusing Air Force generals of fraudulent dealings with their B-36 contractors. Relations eventually grew so tense that Forrestal was hospitalized for depression, committing suicide on 22 May 1949,[3] while the Armed Services Committee described an 'iron wall' – a reference to Churchill's Iron Curtain – existing between the services.[4]

Whether this inter-service rivalry was the real motivation for Keyhoe and McLaughlin's *True* articles, or whether somebody else was using their honest accounts to fight bigger battles for them we can't fully know. These were complex, turbulent times and it's likely that the Air Force, the Navy and the intelligence services were all, to varying degrees, exploiting the flying saucer story to their own ends. And the story would now take a bizarre new turn.

THE MEN BEHIND THE FLYING SAUCERS

Crashed flying saucer stories have always circulated within the UFO community, but were rarely taken seriously until the late 1970s when a number of reports were leaked to UFO researchers via anonymous Air Force sources. The publication of Charles Berlitz and William Moore's *The Roswell Incident* in 1980 was the culmination of that process and the progenitor of the seemingly endless flow of books, magazines, films and merchandise, only a fraction of which were filling the Laughlin conference dealers' room.

However, the origins of the Roswell story don't lie in Roswell itself, but about 350 miles north-west in the small town of Aztec, New Mexico. In 1950 the popular and outspoken *Variety* magazine columnist Frank Scully published a non-fiction book, *Behind the Flying Saucers*, which centred on a bizarre lecture given on 8 March 1950 at the University of Denver, Colorado. In what sounds more like a market research experiment than an academic lecture, ninety science students were asked to attend a presentation on flying saucers by an anonymous lecturer. Word quickly spread around campus and on the day the hall was filled to capacity. In the fifty-minute presentation, the mysterious expert announced that not only were the flying saucers real, but that four of them had landed – not crashed – on Earth and three of these had been captured by the US Air Force.

One of the larger craft, 100 feet across, had landed near Aztec, New Mexico; both the disc and its dead occupants were ferried off to Wright Patterson Air Force Base for examination. In fact this was not a new story: aviation historian Curtis Peebles dates it back to a prank account published by the *Aztec Independent Review* in 1948, which mentioned a saucer crash and little men from Venus.[5] According to Frank Scully, the three captured craft contained the bodies of thirty-four alien beings. These looked much like humans but were smaller in stature, 'of fair complexion' and lacked beards, though some of them had managed to cultivate 'a fine growth resembling peach fuzz'.

The anonymous lecturer described the saucers in some detail. They were:

> quite dissimilar to anything we have designed. There was not a rivet, nor a bolt, nor a screw in any of the ships . . . Their outer construction was of a light metal much resembling aluminum but so hard no application of heat could break it down. The discs . . . had revolving rings of metal, in the center of which were the cabins. . . . the first saucer was capable of maneuvering in any given direction . . . they could be maneuvered to land anywhere. The smallest had a landing gear built like a tricycle of three metal balls, which could revolve in any direction.

Following the Denver lecture, attendees were asked whether they believed the presenter; 60 per cent said they did. Within a few hours, many also found themselves being questioned by Air Force Intelligence officers. Scully notes that a follow-up questionnaire was carried out among the students, and that the number who believed the presentation had fallen from 60 to 50 per cent, which was still considerably higher than the national average who believed that the flying saucers came from Outer Space (about 20 per cent according to Scully). The message from the lecture was clear: exposure to a convincing source of information encouraged even bright college students to believe the improbable.

On 17 March the mystery lecturer was unveiled by the *Denver Post* as Silas Mason Newton, proprietor of the Newton Oil Company, based in Denver, while in his book Scully revealed the source of Newton's saucer information to be 'Dr Gee', a composite name for eight scientists who needed to protect their identities on national security grounds. However, the truth about Newton and Dr Gee turned out to be a little less melodramatic, though no less intriguing.

Gee was exposed as Leo Arnold Julius Gebauer, a man of many aliases and an FBI file thick enough to contain them. Gebauer, who at one time worked in the labs of the Air Research Company in Phoenix Arizona, had come to the FBI's attention in the early 1940s for his outspoken comments about Adolf Hitler, whom he described as a 'swell fellow', and for announcing that President Roosevelt should be shot and replaced with someone like the Fuehrer.[6] Gebauer had told Newton that he worked with government agencies retrieving technology from downed flying saucers, including the one at Aztec. Whether or not Newton actually believed him is unclear, but this didn't stop him from promoting Gebauer's tales to the students in Denver.

In his diaries Newton wrote that after his identity was revealed by the *Denver Post* he was approached by two members of a 'highly secret US Government entity' who told him that they knew his UFO crash story was a hoax, but that he should continue to tell it. If he did then 'they and the people they worked for would look out for me [Newton] and for Leo [Gebauer]'.[7] Were these mystery men figments of Newton's devious imagination, Air Force Intelligence agents, FBI or CIA officers or, perhaps, Navy men out to cause more trouble for the US Air Force? We can only wonder, but they got what they wanted. Scully's book was rushed out in 1950 and sold around 60,000 copies, making it a bestseller of its day and further cementing the details of the flying saucer myth in the American imagination. Newton had done his job well. And perhaps Newton's mysterious government men also upheld their side of the arrangement: when, in 1952, Newton and Gebauer were convicted of fraud for trying to sell advanced mining equipment based on back-engineered alien technology, both received only suspended sentences.

Despite the success of *Behind the Flying Saucers*, the Aztec saucer crash would be almost entirely forgotten within a few years until, nearly three decades later, its key elements – the craft, the dead pilots, the Air Force recovery operation and the back-engineering programme at Wright Patterson – would form the basis of the Roswell story. In the early 1980s, Aztec itself became part of AFOSI's disinformation campaign against the UFO community, leading to the publication of another book[8] promoting the crash as a genuine event.

So, in the span of half a century, a story that began life as a newspaper prank became a reality, was dismissed as a hoax, was resurrected and promoted by the US Air Force and UFO researchers in the 1980s before finally being laid to rest (we hope) as a non-story in the early twenty-first century. If nothing else, this proves that flying saucers are highly recyclable.

Folklorists have a word for the process whereby folktales bleed into reality; they call it 'ostension'. But when these tales are given a kick start by the intelligence agencies, I think we can simply call it deception. Looking back at the UFO's first decade, it's clear to see that the military and civilian intelligence agencies acted as midwives in the birthing of the UFO myth. There are signs of intelligence 'dirty tricks' at work in the Maury Island case and, if we believe Silas Newton's diaries, then at least one intelligence agency was complicit with the seeding of the crashed UFO and occupant stories surrounding Aztec. It was a pattern that would be repeated over and over again in the coming decades, and UFOs were just one of the tools that the intelligence tricksters used to do their dirty work.

SPOOKS, GHOSTS, VAMPIRES AND ALIENS

With a few strokes of his pen, on 26 July 1947 President Harry Truman separated the US Air Force from the Army and transformed the wartime Office of Strategic Services into the Central Intelligence Agency. Within a few years the CIA, originally intended as an organizing body for the three military intelligence groups, had become the great temple of the cult of intelligence, a cult that, it sometimes seemed, sought not only to rule the world, but to control the hearts and minds of every person in it. And for three decades it seemed that nobody could control the CIA.

In their book *The CIA and the Cult of Intelligence*, John Marks and Victor Marchetti make clear that for many years the CIA, as an institution, was not only out of control, but beyond it:

> The CIA . . . penetrates and manipulates private institutions, and creates its own organizations (called 'proprietaries') when necessary. It recruits agents and mercenaries; it bribes and blackmails foreign officials to carry out its most unsavoury tasks. It does whatever is required to achieve its goals, without any consideration of the ethics involved or the moral consequences of its actions . . . Its practices are hidden behind arcane and antiquated legalisms which prevent the public and even Congress from knowing what the mysterious agency is doing – or why.'[9]

There is perhaps no more explicit example of CIA attitudes to secrecy than the response of its director Richard Helms to the exposure of MK-ULTRA, the Agency's long-running programme that used drugs and hypnosis to conduct 'brainwashing' and mind-control experiments. When, in the mid-1970s, a Congressional investigation demanded to see all the Agency's MK-ULTRA documentation, Helms ordered it to be destroyed; he almost succeeded, though the little that remained was still enough to force a radical overhaul of CIA practices.

When it came to secret warfare – espionage and counter-espionage, intelligence and counter-intelligence, psychological operations, disinformation and covert action – nothing was true and everything was permitted, just as long as nobody found out. Sometimes this required the manipulation of small armies, sometimes it required the development and employment of arcane technologies, but sometimes, as is the case with a good magic trick, a little suggestion was all that was required to make an impact.

Air Force Colonel Edward Lansdale was an early master of the art of psychological warfare. One of America's most feared and respected 'Cold Warriors', Lansdale was a former advertising man turned intelligence specialist, a legend in his own lifetime and the inspiration for Alden Pyle in Graham Greene's *The Quiet American*. His years in the advertising game served Lansdale extremely well in the intelligence world. Here he coupled his understanding of the power of perception and presentation with a firm conviction that America's well-being required dominance of the Third World, a goal that was to be achieved by winning over the 'hearts and minds' of the population and creating a state of economic dependency on the US.

Having fought in the Philippines during the Second World War, in the early 1950s Lansdale was recruited by the Agency to assist the Philippine defence minister in his battle against the local communist guerrilla insurgents, the Huks. As part of his campaign, Lansdale set up the Filipino Civil Affairs Office as a base for psychological warfare operations (PSYOPS). Like market researchers at an ad agency, Lansdale's team tried to get inside the minds of the local communities, to find out how they lived, what they most hoped for and, of course, what they most feared.

One PSYOPS project involved flying a small plane over Huk territory, hidden in thick cloud cover. The aircraft used megaphones to broadcast the 'voice of God', which warned villagers who sheltered or fed insurgents that they would be cursed. Another operation exploited rural superstitions surrounding the *aswang*, a vampire from Philippine mythology. Lansdale's team seeded rumours that an *aswang* lived in a region occupied by the Huk. Word of the bloodsucking menace soon spread among the guerrillas and their supporters until, one day, their worst fears were confirmed: a guerrilla was found dead with puncture marks on his throat and the blood drained from his body. Rather than being a victim of the *aswang*, however, the unlucky Huk had been ambushed by Lansdale's team, killed and hung from a tree until he was drained of blood, then laid out to be discovered by his colleagues. To the other Huk, this confirmed the *aswang* stories and they fled the area in terror. By 1953 the attempted communist insurgency had been successfully quashed; afterwards Lansdale became one of the first operatives to enter Vietnam, paving the way for the American incursion there, and following that, played a key role in Operation Mongoose, which made eight unsuccessful attempts on Fidel Castro's life.

Local superstitions were also exploited during the Vietnam War, where the Army's 6th PSYOPs Battalion regularly broadcast an audio recording called 'The Wailing Soul' through speakers mounted on backpacks or helicopters. Preying on Vietnamese traditions of the unquiet dead, the tape contained a conversation between a little girl and the wandering soul of her dead father, who had been killed while fighting the Americans. The recording, which made heavy use of eerie reverb effects and traditional Vietnamese funeral music, was so effective that it also spooked American soldiers patrolling the jungle at night.

Lansdale's *aswang* and the wandering soul were just two of countless psychological deception operations carried out during the hot years of the Cold War. Tom Braden, former head of the International Organizations Division of the Directorate of Plans (now the National Clandestine Service), which oversaw most of the CIA's PSYOPs, covert action and propaganda work, wrote in 1973 that there were 'so many CIA projects at the height of the Cold War that it was almost impossible for a man to keep them in balance'.[10]

In the fight against communism, maintaining a firm but gentle grasp on hearts and minds at home – the proverbial iron fist inside a velvet glove – was as important as winning them over abroad. Although the National Security Act expressly forbade the CIA from conducting activities on American soil, it seemed to have no trouble finding ways to do so, setting up a veritable empire of false companies – nicknamed 'Delawares' after the state in which they were registered – and employing 'quiet channels', companies and institutions who were on the right side, to get their people into key positions on newspapers, magazines, TV and radio stations, businesses and grass roots organizations across the nation. While the CIA worked on the ground, the bigger picture was shaped by an even more secretive organization, about which little was known until almost fifty years after its dissolution.

The Psychological Strategy Board (PSB) was signed into existence by Harry Truman in 1951, tasked with co-ordinating psychological operations at home and abroad, and ensuring that America and Americans looked, sounded and thought right. If this sounds Orwellian, then that's because it was: even the contents of its first strategy paper are still classified, though traces of it can be found referenced in other documents. According to one, the PSB's role was to develop 'a machinery' to promote 'the American way of life', and to counter 'doctrines hostile to American objectives'. To do so they would take in 'all fields of intellectual interests, from anthropology and artistic creations to sociology and scientific methodology'.[11]

In May 1952 the PSB took over Packet, the CIA's psychological warfare programme, aimed at persuading foreign leaders that the American way was superior to anyone else's way, particularly the Russians'. Maintaining America's charisma abroad required the control, procurement and production of everything from scholarly 'seminars, symposia, special tomes, learned journals [and] libraries', to church services, comic books, 'folk songs, folk lore, folk tales and itinerant storytellers'. The PSB's message was broadcast over TV and radio, and from ships and aircraft; even the use of 'three dimensional moving images' was considered for added realism[12] (American cinemas were enjoying a 3D boom at the time).

Charles Douglas 'CD' Jackson, who would be a close adviser to President Eisenhower after his election, was a key strategist for the

PSB. Like Edward Lansdale, Jackson was a former advertising executive and magazine publisher (at *Time Life* and then *Fortune*) turned intelligence specialist. A champion of American values, Jackson was regarded as the most influential member of the invisible government that shaped America's image after the war, and he had powerful friends in the arts, such as Henry Luce who ran the *Time Life* empire and Hollywood mogul Darryl Zanuck. Where they didn't already have a strong grip Jackson and the PSB created one, gaining influence over publishers, newspapers, TV and radio broadcasters, artists and art organizations, orchestras and small but influential magazines such as *Encounter* and the *Partisan Review*. PSB influence generally required just a friendly word in the right ear, but sometimes it took money and occasionally it necessitated total control over a target body to get the message out: if opinions needed to be formed, the Board was there to do it.

At times the PSB could be bold and direct: in 1952 Jackson noted that work was progressing well on a *LIFE* magazine article by Gordon Dean, chairman of the Atomic Energy Commission, that would 'remove the guilt complex from America on the use of the A-bomb'.[13] More usually a little subtlety was required; as Eisenhower unrolled his Atoms for Peace programme in 1954 another memo from Jackson discussed the propaganda potential of the President's plan to build an atomic power plant in Berlin. Jackson pointed out that it wouldn't be necessary actually to construct the plant. By fencing off an area of rubble, assigning guards to it and putting up mysterious signs, they could create a rumour wave as powerful in its effect as if they had erected the real thing.

The Psychological Strategy Board was rebranded in 1953 as the more oblique-sounding Operations Coordinating Board, but the wheels set in motion by Jackson and his team continued to spin throughout the 1960s, curbed only after Senator Frank Church led his 1973 investigation into CIA activities. Following this unwelcome exposure the Agency was forced to out 400 employees and agents working within the media, an estimate generally regarded as being 'on the low side'.[14]

So what's this got to do with UFOs? Possibly everything: recognizing how the PSB, the CIA and America's political and intelligence elite worked and thought during the early 1950s is critical to understanding what happened when the CIA turned its mind to the flying saucer problem.

SIX

WASHINGTON VERSUS THE FLYING SAUCERS

'Symbols should convey the Line of Persuasion. They must convey a preconceived notion already developed by the deception target . . . Sport anglers do the same by applying scents, motion, and color to indicate the lure is an easy meal.'

'A Primer For Deception Analysis: Psychological Operations' Target Audience Analysis', Lt Col. Rieka Stroh and Major Jason Wendell, Iosphere, *Fall 2007*

In early 1952 CIA Director Walter B. Smith wrote to Raymond Allen, director of the Psychological Strategy Board:

> I am today transmitting to the National Security Council a proposal in which it is concluded that the problems associated with unidentified flying objects appear to have implications for psychological warfare as well as for intelligence and operations. I suggest that we discuss at an early board meeting the possible offensive and defensive utilization of these phenomena for psychological warfare purposes.[1]

Smith was responding to the dramatic growth in public interest in flying saucers during the last part of 1951, which had led to a sudden surge in witness reports, many from within the armed forces. At the same time the US Air Force's Project Grudge team had become rather too good at their job of playing down the saucer problem, to the extent that they were ignoring some genuinely puzzling incidents. A sighting by pilots and radar operators over Fort Monmouth Army Base in New Jersey that September was the final straw for Air Force high command, and in March 1952 Grudge was replaced by Project Blue Book. This was just in time for what would be a bumper year for flying saucers, with 886 reports received between June and October of 1952, coming in at fifty a day during the summer peak. Blue Book head Captain Edward Ruppelt – who coined the term Unidentified Flying Object – noted that this was 149 more than had been received in total by the Air Force since 1947.

Following the Fort Monmouth sighting, which Ruppelt thought was a balloon, the Air Force issued JANAP 146(B), an expansion of a directive issued to all the armed forces, instructing them to report sightings of unknown aircraft to the Secretary of Defense (who then passed on the reports to the CIA), Air Defense Command, and the nearest US military base. The unauthorized release of information about a UFO incident was now a crime, with a penalty of up to ten years in prison and a $10,000 fine. With the Soviets watching America's every move, UFOs – and that included clandestine balloon and missile launches and test flights of aircraft under development – were a growing intelligence and security problem, and one that needed to be contained.

Walter Smith's concerns proved to be uncannily prescient. The UFO situation reached an embarrassing, potentially catastrophic climax on two nights in July 1952, when a number of unidentified objects blipped on to radar screens at Washington DC National Airport. At close to midnight on the first night, 19/20 July, seven objects were tracked fifteen miles from the capital city, gradually homing in on the White House at about 100 m.p.h. A bright, orange ball of light was seen from nearby Andrews Air Force Base, making 'a kind of circular movement' according to an airman on the scene, before taking off at 'an incredible speed' and disappearing.[2] Six bright white, fast-moving lights were also spotted by the pilot of a passenger jet flying in the area.

Sightings and radar tracking of unidentifieds continued until 3 a.m., when two interceptors flew in to try to get a closer look, at which point the remaining UFOs vanished from the skies and from radar. They reappeared as soon as the jets had returned to base, having run low on fuel, leading Harry Barnes, senior Air Traffic Controller for the Civil Aeronautics Administration, to suspect that the UFOs, whatever they were, were listening in on radio communications and planning their actions accordingly. Adding to his frustration, Barnes's attempts to interest senior Air Force officials in the incident seemed to fall on deaf ears. Creating further grounds for suspicion that someone in the Air Force knew what was going on, Blue Book's Edward Ruppelt didn't hear about the case until he read about it on the front page of the newspapers two days later.

An intrusion by unknown aircraft into US airspace might seem impossible to us now, but we can imagine the furore that it must have caused, coming only eleven years after Pearl Harbor, a wound still fresh in America's military memory; and in 1952 the threat posed by such an incursion was far greater than it was in 1941. The Soviets had by now detonated three atomic bombs: any one of the seven UFOs over Washington that night could have been a Russian bomber loaded with their own Fat Man or Little Boy. Then, on 26 July, the UFOs came back. This time twelve were spotted on radar, again flying at a not particularly impressive 100 m.p.h. As before there were sightings of lights from the air and from the ground and, once more, two jets were scrambled. One of the pilots chased four white 'glows' that suddenly 'shot toward him and clustered around his plane'[3] but the UFOs remained as elusive as ever.

Another media flurry followed, leading to an Air Force press conference at the Pentagon, its largest since the Second World War. In his 1956 UFO memoir, *The Report on Unidentified Flying Objects*,

Ruppelt describes the scene as chaotic, with General John Samford of Air Force Intelligence doing his best to be noncommittal about the sightings and focusing on calming fears that they were guided missiles or secret American aircraft. When asked directly whether the objects had been US secret weapons, Samford gave the oblique and enigmatic response 'we have nothing that has no mass and unlimited power'. Then came Captain Roy James, a radar specialist from the Air Technical Intelligence Center (ATIC) at Wright Patterson, who pointed out that at least some of the radar returns were the result of a temperature inversion, a layer of warm, moist air on top of cool air on the ground, which had caused radars to pick up a steamboat and other large objects at ground level. Ruppelt and others were unconvinced by the explanation[4] but the press lapped it up.

The two overflights were eerily similar to events portrayed in the previous year's hit film *The Day the Earth Stood Still*, in which the benevolent humanoid alien Klaatu's flying saucer caused panic as it first approached, then landed in Washington DC. As in the film, the real-life Washington sightings made front-page news all over the country, instigating a massive surge of sightings and more work for Ruppelt and Blue Book. Nationwide reports flooded in to the Air Force, reaching a record 536 for the month of July alone and threatening to swamp internal communications and overwhelm the news media with sensational accounts. Over the Atlantic, the deluge piqued the interest of British PM Winston Churchill, who asked in a memo to his advisers: 'What does all this stuff about flying saucers amount to? What can it mean? What is the truth?'[5]

All this activity made the CIA jumpy. Something needed to be done: it was time for the Agency to enter the UFO business. The CIA's UFO investigation drew in its Office of Current Intelligence, the Office of Scientific Intelligence and the Weapons and Equipment Division. In August 1952, its representatives gathered for a number of highly secret meetings with its Air Technical

Intelligence Center counterparts at Wright Patterson. It was critically important that CIA involvement with the UFO subject be kept out of an increasingly wary public eye: the first whispers of a 'conspiracy' and 'cover-up' by a 'silence group' were already doing the rounds, thanks to Donald Keyhoe, whose 1949 *True* article had become a bestselling book, *Flying Saucers are Real*. The knowledge that the CIA were getting into UFOs could only confirm and amplify these suspicions.

The briefing paper for the CIA investigation provides a good sense of the way that the Agency, and thus the nation's custodians, viewed both the UFO phenomenon and the rest of the world. It also demonstrates to us now that the problems raised by the issue over half a century ago have changed very little. The brief first considered the four main operational theories for UFOs: that the objects were classified US aircraft; that they were Russian aircraft; that they were extraterrestrial in origin, and, finally, that they could be explained as misperceptions of known aircraft and natural phenomena. It noted that CIA officers had pursued the first issue, of secret aircraft, to the very highest authority, and were satisfied that no currently operational project was responsible for the sightings. They ought to have known, as within three years the Agency would be flying the U-2, the most highly secret aircraft of its time, and one that would be responsible for more than its fair share of UFO sightings. They also pointed out that, even if the Air Force was lying to them, the evidence weighed against this suggestion: why would the US Air Force risk scrambling its own jets against its most precious new toys, and why would they take the incredible risk of flying them in public airspace over the capital?

A similar question could be asked of the logic behind Soviet overflights. While the CIA knew that, as in America, Russian engineers had looked into the possibilities of elliptical and delta-wing aircraft designs, they were unaware of any technological leaps that would have got them off the ground, even if the Russkies were crazy enough to fly them over the enemy's capital. Nor was there any logic to the pattern of the overflights that might suggest a reconnaissance project. Another theory, although one that was 'totally unsupported', was that the Russians were flying balloons over the US and using press reports to gauge their flight paths. Although this was also considered highly unlikely, there was a precedent for it: overflights by Japanese 'fugu' fire bomb balloons had resulted in US civilian deaths as late as 1945. As to 'the man from Mars', the CIA admitted that 'intelligent life may exist elsewhere' but found no astronomical evidence to support the theory that it was visiting Earth, nor, once again, did the pattern of sightings make any sense from an operational perspective, the same conclusion reached by RAND's James Lipp four years previously.

This left only the fourth option – that the UFOs were the result of a range of misperceptions – as the most likely answer, and the one that, since the closure of Project Grudge, had become the official Air Force line. With this in mind, the briefing paper noted that the people making the reports were, for the most part, earnest in their intentions: the misidentified objects were almost always seen against the sky, providing no reference point from which to measure size, speed, distance or motion. A range of psychological factors were also recognized as underpinning many UFO sightings: the influence of media reports (the CIA use the Orwellian term 'mental conditioning'), the possible desire for publicity resulting in the embroidering of facts or complete fabrication, and the emotional response of individuals to encounters with the unknown.

On 24 September 1952, H. Marshall Chadwell, the CIA's assistant director for Scientific Intelligence, sent a summary report of the ATIC meetings to Director Walter Smith, outlining the conclusions drawn from the sessions. It is worth reprinting almost in full:

> the flying saucer situation contains two elements of danger which, in a situation of tension, have national security implications. These are:

a) Psychological – with worldwide sightings reported, it was found that, up to the time of the investigation, there had been in the Soviet press no report or comment, even satirical, on flying saucers . . . With a State-controlled press, this could result only from an official policy decision. The question, therefore, arises as to whether or not these sightings:

1) Could be controlled
2) Could be predicted
3) Could be used from a psychological warfare point of view, either offensively or defensively.

The public concern with the phenomena, which is reflected both in the United States press and in the pressure of inquiry upon the Air Force, indicates that a fair proportion of our population is mentally conditioned to the acceptance of the incredible. In this fact lies the potential for the touching off of mass hysteria and panic.

b) Air Vulnerability – the United States Air Warning System will undoubtedly always depend upon a combination of radar screening and visual observation. The USSR is credited with the present capability of delivering an air attack against the United States . . . At any moment of attack, we are now in a position where we cannot, on an instant basis, distinguish hardware from phantom, and as tension mounts we will run the increasing risk of false alerts and the even greater danger of falsely identifying the real as phantom.

As well as instigating an investigation into what the Soviets knew about flying saucers, Chadwell concluded that a 'study should be instituted to determine what, if any, utilization could be made of these phenomena by United States psychological warfare planners and what, if any, defenses should be planned in anticipation of Soviet attempts to utilize them'. His final recommendation was that they should manage public perceptions of the phenomena 'to minimize the risk of panic'.

Walter Smith didn't need much more convincing, and in January 1953 the CIA convened a secret panel consisting of nuclear physicists, radar and rocketry experts, other Air Force personnel and an astronomer. Headed by Dr Howard Percy Robertson, director of the Pentagon's Weapons Systems Evaluations Group, the group spent four days with very long lunch breaks listening to UFO reports, watching films of unexplained objects, and seeking possible explanations for the phenomenon. Their conclusions, which were not fully revealed to the public until 1966, effectively echoed the concerns of Chadwell's earlier report.

The Robertson Panel Report recommended that the military focused on training its personnel to recognize unusually illuminated man-made objects and natural phenomena, including meteors, fireballs, mirages and clouds, both visually and on radar. 'Such training', it noted, 'should result in a marked reduction in reports caused by misidentification and resultant confusion.' As far as the general public were concerned, a programme of 'debunking' would be set in place to reduce their interest in the subject, and diminish 'their susceptibility to clever hostile propaganda' from the Soviets. 'As in the case of conjuring tricks,' it noted, 'there is much less stimulation if the "secret" is known.' The Panel had some interesting suggestions as to how this education might be carried out: they recommended using Walt Disney cartoons and the Jam Handy Co., which made Second World War training films, as well as the Navy's own Special Devices Center (now the Office of Naval Research) to train in aircraft identification.

Psychological monitoring of the population was another key consideration. Panel members would have been aware of an extreme case of UFO nerves from Quito, Ecuador, on 12 February 1949, when a panic triggered by a radio broadcast of *War of the Worlds* led to riots that were quelled only after tanks took to the city streets, resulting in twenty deaths. It was 'felt strongly' that psychologists should advise on the programme – Hadley Cantril, who had written a study[6] of the panic in the US surrounding Orson Welles's original 1938 *War of the Worlds* radio play was mentioned – along with 'someone familiar with mass communications techniques, perhaps an advertising expert'.

The Robertson Report also recommended that civilian UFO groups be monitored, 'because of their potentially great influence on mass thinking if widespread sightings should occur. [Their] apparent irresponsibility and the possible use of such groups for subversive purposes should be kept in mind.' For the next two decades, one of the groups mentioned by name, the Aerial Phenomena Research Organization (APRO) of Tucson, Arizona, found itself under close scrutiny by the intelligence services.

In conclusion the Report found that while the UFOs themselves seemed to present no 'direct physical threat to national security', the reporting of them did: 'clogging . . . channels of communication by irrelevant reports' and creating a cry wolf situation that could create so many false alarms that genuine hostile actions might be ignored. What's more, the general interest in the subject threatened to inculcate 'a morbid national psychology in which skillful hostile propaganda could induce hysterical behavior and harmful distrust of duly constituted authority'.

Flying saucers could make a rebel out of you, or worse, a Communist. The national security agencies were, therefore, to 'take immediate steps to strip the Unidentified Flying Objects of the special status they have been given and the aura of mystery they have unfortunately acquired'. As Curtis Peebles observes, 'the Robertson Report was not really about flying saucers, it was about Pearl Harbor . . . the US was haunted by the specter of a surprise Soviet nuclear attack.'[7]

To what extent the CIA and the US Air Force put its recommendations into action is not always clear, but one outspoken scientist, Leon Davidson, now forgotten by all but the most committed UFO buffs, thought that he had the answers.

CIA+ECM=UFOS

A child of the inter-war generation, Leon Davidson had always been something of a scientific prodigy. By the age of thirteen he had declared himself a chemical engineer, and a few years later he would be plucked from his PhD course at the engineering school of Columbia University to work on the Manhattan Project. He eventually became a supervising engineer at the Los Alamos laboratories and worked for many years on computer systems for the nuclear industry before developing an early interest in touch-tone telephone technology.

Like many scientists working within what Eisenhower would call the 'military industrial complex', Davidson became fascinated with the UFO problem. Soon after starting work at Los Alamos in 1949, he joined the labs' Astrophysical Association, an in-house flying saucer group investigating a wave of strange green fireball sightings taking place around New Mexico. Davidson was convinced that the fireballs were part of a covert upper atmosphere rocket research programme. No official explanation was forthcoming but he gradually came to believe that secret military tests lay behind most UFO incidents. A front-page *Washington Post* report about the capital's UFO overflights refers to Davidson as a scientist with a special interest in flying saucers:

> Davidson, whose study of saucers is impressively detailed and scientific, said he believes the [UFOs] are American 'aviation products' – probably 'circular flying wings,' using new type jet engines that permit rapid acceleration and relatively low speeds. He believes, they are either a 'new fighter', guided missiles, or piloted guided missiles. He cited some of the recent jet fighters, including the Navy's new F-4-D, which has a radical 'bat-wing', as examples of what the objects might resemble.[8]

By 1959 what had started as a suspicion for Davidson was beginning to seem like a certainty, and he suggested that the

Washington UFOs had been a deliberately staged test of advanced electronic counter-measure (ECM) technology. In the splendidly titled essay 'ECM+CIA=UFO' from the February/March 1959 edition of the ufological insiders' newsletter *Saucer News*, he described the basic ECM technology available to the US Air Force by 1950:

> A 'black box' in our bombers would pick up the enemy's radar impulses; amplify and modify them; and send them back, drowning out the normal radar return from the bomber. The modification could be a change in timing or phase and could cause the 'blip' on the radar screen to have an incorrect range, speed, or heading.[9]

The most primitive form of ECM, called 'window' or chaff, wasn't really electronic at all. First used by the British during a devastating bombing raid on Hamburg in July 1943, it required crew members to throw packets of aluminium strips out of their aircraft. Cut to half the wavelength of the German Würzburg radars (53–4 centimetres), the billowing metallic clouds overloaded enemy radars with false echoes. As the war progressed, more complex electronic systems were installed in military vehicles to jam radars and radios at specific wavelengths and frequencies.

These were all methods of obfuscation; what Davidson was talking about is a little more sophisticated: deception. He identified a wartime incident from the Nansei Shoto archipelago in the South Pacific as the origin for this new procedure. It was April 1945, nerve-wracking times for the Allies as they prepared to invade Okinawa in one of the last assaults of the Second World War. All ships in the region dreaded being targeted by Japanese kamikaze squadrons and any blips on their radar screens would send all hands scrambling on deck to fight back. Sometimes, however, radar blips would appear with no aircraft to account for them. These phantom radar returns, known as the 'galloping ghosts', appeared repeatedly on the assembled Nansei Shoto ships' radar screens. At least some of these ghosts were caused by flocks of birds: individual pelicans – later to be blamed by some for Kenneth Arnold's saucer sighting – could even be mistaken for solo aircraft. Navy scientists also suspected that the close proximity of so many powerful naval radars had caused some of the ghosts, and realized that being able to produce radar phantoms at will would be an extremely useful tool for bamboozling the enemy.

A March 1957 article from *Aviation Research and Development* magazine discussed how this ghosting technology had been improved and was now being released into the civilian domain:

> A new radar moving target simulator system which generates a display of up to 6 individual targets on any standard radar indicator has been developed . . . to train radar operators . . . and for in-flight testing of airborne early-warning personnel . . . Target positions, paths, and velocities can . . . simulate . . . realistic flight paths . . . Speeds up to 10,000 knots [about 11,500 m.p.h.] are easily generated . . . The target can be made to turn left or right . . . For each target there is . . . adjustment to provide a realistic scope presentation.'[10]

Davidson recognized this description as being close to what was spotted on radar over Washington in July 1952 – and he knew just who had been playing with them:

> Since 1951, the CIA has caused or sponsored saucer sightings for its own purposes. By shrewd psychological manipulation, a series of 'normal' events has been served up so as to appear as quite convincing evidence of extraterrestrial UFOs . . . [including] military use of ECM on a classified basis unknown to the radar observers who were involved.[11]

Was Davidson right? It may not have been the CIA, but it seems likely that *someone* was using this technology to test pilots and radar operators. A 1957 incident that took place in the UK appears to be a classic case of radar spoofing at the expense of a terrified American pilot. Twenty-five-year-old Lieutenant Milton Torres was based at RAF Manston in Kent, then an outpost for America's Strategic Air Command in Europe. On 20 May he received the order to scramble his F-86D Sabre jet in pursuit of a large aircraft the size of a B-52 bomber (approximately 160 feet long and 180 feet wide), picked up on radar about fifteen miles away. Torres was given the order to arm his weapons and fire on sight, something that no airman would expect to have to do over the Kent countryside, except at a time of war. As he feared, he was informed that the aircraft was hostile and probably Russian.

Torres and a wingman in another Sabre flew to 32,000 feet and hurtled at Mach 0.92 (about 700 m.p.h.) towards the huge object – according to Torres, registering as the size of an aircraft carrier, yet zipping about on his radar screen like an insect. He was ready to fire a full salvo of twenty-four rockets at the intruder, yet neither he nor his wingman could see a visual target. Was the aircraft invisible? Suddenly the radar signature disappeared and the Sabres were called back to base.

The next day Torres was visited by a trenchcoated American who claimed to be from the National Security Agency. The mystery man warned Torres that if he ever wanted to fly again he would keep his mouth shut. And Torres did, for thirty years.[12] Torres's story sounds like a textbook example of the radar spoofing described by Davidson (the annals of UFO history are full of similar accounts). Whether the mysterious American was really from the NSA we can't know: both that agency and the CIA had reason to be interested in the technology, and both had a habit of disguising their contact teams as members of other agencies.

The CIA and the NSA also worked together: by the start of the 1960s they had instigated a project known as Palladium, designed to provide the Americans with electrical (ELINT), communications (COMINT) and signals (SIGINT) intelligence from Soviet aircraft, ships, submarines, ground radars and missile batteries. In the first years of the Cold War this intelligence was gathered the hard way, by pilots known as 'ravens' flying their aircraft on dangerous 'ferret' missions. The ravens used their aircraft to probe the outer perimeters of Soviet airspace, tripping its defences to collect as much data as possible from ground radars and communications systems.

Palladium presented them with a much safer way to gather data. The technology allowed the CIA to create ghost aircraft to be detected on Soviet radars, while the NSA monitored the way in which the phantoms were received, tracked and transmitted. These ghost aircraft could be built to order in any shape and size, and could fly at any speed or altitude.

Former CIA signals specialist Eugene Poteat describes a complex operation during the Cuban Missile Crisis that used both the Palladium system and submarine-launched metallic spheres on parachutes to confuse Cuban radars. Poteat's CIA team flew a radar ghost into Cuban airspace, prompting fighter planes to be scrambled to intercept the 'intruder'. Using the Palladium system's controls the CIA kept their phantom aircraft just ahead of the Cuban fighters, waiting for the right moment. When the NSA team heard that the Cuban pilot was about to make a pass to shoot down their ghost plane 'we all had the same idea at the same instant. The engineer moved his finger to the switch, I nodded yes and he switched off the Palladium system.'[13]

Edward Landsale's *aswang* – the vampire from Philippine mythology – was now an airplane. It seems reasonable to suggest that a number of early military UFO incidents involving radar were deliberately spoofed, both to test the responses of radar operators and pilots to such anomalies, and to test the technique's potential in psychological warfare scenarios. Leon Davidson, however, took things one giant step further. Palladium, he suggested, was just one tool in achieving the government's ultimate aim: the transformation of unidentified flying objects into extraterrestrial spacecraft, and the staging of an alien invasion, just as Bernard Newman had described in *The Flying Saucer*.

MANUFACTURING THE MYTH

The July 1952 flying saucer event over Washington was a pivotal moment in UFO history. It prompted exactly the sort of confusion that the CIA had warned the Psychological Strategy Board (PSB) about and gave the Agency a perfect reason to get involved in the UFO business. It also generated fresh news headlines all over the world, just as public interest in the subject was peaking in the UK.

But had America already been primed for an alien invasion? In April 1952, *LIFE*, America's most popular magazine, ran an article entitled 'Have We Visitors from Space?' The issue featured a pouting young Marilyn Monroe on the cover, making it irresistible to any red-blooded American male. 'The Air Force', the article begins, 'is now ready to concede that many saucer and fireball sightings still defy explanation; here *LIFE* offers some scientific evidence that there is a real case for interplanetary saucers.' The article climaxes by insisting, in bold type that they are *not* a natural phenomenon, *not* American or Russian secret aircraft, *not* balloons and *not* psychological in origin. Therefore they must be from Outer Space.

Its authors, H.B. Darrach Jr. and Robert Ginna, had spent a year in consultation with the US Air Force, who were still supposed to be keeping a lid on the flying saucer subject, so the determinedly pro-ET tone seems something of a surprise. Why, Davidson wondered, would one of the most respectable magazines in the US run an article like that unless the Air Force wanted it to? According to Ruppelt, Ginna had been speaking to some very senior Air Force men, whose opinions were reflected in the piece. But was it Air Force strategy or, given *Time Life* proprietor Henry Luce's close relationship with the CIA and the PSB, somebody else's game plan that Ginna and Darrach were following?

The *LIFE* article promoted the ET hypothesis while giving flying saucer studies a further boost of respectability, and no doubt contributed to the deluge of press reports that the phenomenon was generating: according to Ruppelt, 148 US newspapers carried more than 16,000 items about UFOs in the first six months of 1952.

Was somebody deliberately drumming up saucer hysteria as a prelude to July's spectacular Washington overflights? Davidson was convinced that the Washington wave had been a demonstration, a case of radar spoofing on a grand scale. With the benefit of a little perspective the idea, while still paranoid, isn't quite as crazy as it sounds. There's no question that the CIA had taken a keen interest in the UFO phenomenon by 1952, and it seems perfectly reasonable for them to have done so. By 1957, the technology to create and control accurate ghost aircraft on radar was available to anyone who could pay for it, based on radar spoofing principles understood as far back as 1945. A statement made by General Samford to the *New York Times* soon after the Washington incident might also be read as hinting at deliberate radar manipulation: 'We are learning more and more about radar . . . [which is] capable of playing tricks for which it was not designed'.[14] Davidson also notes that during the month the overflight took place, the Air Force interceptors tasked with protecting the capital were moved from their usual home at Andrews AFB, four miles from DC, to New Castle, Delaware, ninety miles away. This was due to alleged runway repairs at Andrews and considerably delayed the jets' arrival on the scene.

However, the clearest hint that the Washington sightings were no accident was given to Project Blue Book's Edward Ruppelt a few days before events over DC kicked off in July. Following a series of sightings of strange lights by airline crews, Ruppelt wrote that he and a scientist 'from an agency I can't name' (Davidson assumed this was the CIA) had a two-hour discussion about UFOs, at the end of which the scientist made a 'prediction': 'Within the next few days . . . they're going to blow up and you're going to have the granddaddy of all UFO sightings . . . in Washington or New York . . . probably Washington.'[15]

A few days later, it happened – just as the scientist had said it would. As Ruppelt complains in his book, Air Force Intelligence were the last to know about the Washington event and, as we saw earlier, he himself found out about it via the newspapers only two days later. When Ruppelt then tried to get to Washington to investigate at first hand he found that he couldn't get a staff car: 'Every time we would start to leave,' he wrote, 'something more pressing would come up.'[16] So the Air Force's chief UFO investigator stayed put at Wright Patterson and missed being on site for probably the most dramatic event in flying saucer history. Ruppelt later claimed that compiling his report on the Washington overflights took him a year, but could have been done in a day had he been able to get to Washington on time. It was as if somebody didn't want him to do his job. 'I have no idea what the Air Force is doing,' he told *LIFE* journalist Robert Ginna soon afterwards.

WHAT STORK KNEW

If the Washington flap was a set-up, then who was responsible and what was its purpose? A controversial possibility lies between the lines of a memo sent from Dr Howard Clinton Cross to Edward Ruppelt, via Colonel Miles Gol on 9 January 1953.[17] Cross was a metallurgist working at the Battelle Memorial Institute, a private research body specializing in material sciences; at the time, Battelle was processing all the US Air Force's UFO data under the codename Project Stork. Miles Gol was chief of analysis for the US Air Forces's technology transfer division. The involvement of Cross, a metallurgist, Gol, a hardware specialist, and Ruppelt, a ufologist, makes clear that whatever the Air Force thought about UFOs, they were considered relevant to their force's technological concerns.

The memo, classified 'Secret', points out that the CIA's Robertson Panel was due to meet in less than a week's time and that Project Stork and the Air Force's Air Technological Intelligence Center (ATIC) should discuss beforehand 'what can and what cannot be discussed at the meeting'. This suggests that the Air Force was prepared to withhold information from the CIA about UFOs. But what information? It could have been that Project Stork still had no satisfactory answer to the UFO question. Repeating what other internal US Air Force reports had found, Cross writes: 'Experience to date on our study of unidentified flying objects shows that there is a distinct lack of reliable data with which to work.'

Was Cross too embarassed to tell the CIA that the Air Force didn't know what was going on? Perhaps, but his next suggestion has an altogether more conspiratorial tone. As part of Project Stork, Cross recommends that 'a controlled experiment be set up by which reliable physical data can be obtained'. Stork planned to set up observation platforms in areas susceptible to UFO sightings, from where they could make detailed recordings of weather patterns, radar returns and any unusual visual phenomena that could be misinterpreted as UFOs, including balloons, aircraft and rocket tests. 'Many different types of aerial activity should be secretly and purposefully scheduled within the area', writes Cross; '. . . it can be assumed that there would be a steady flow of reports from ordinary civilian observers, in addition to those by military or other official observers.' Cross is proposing that UFO incidents be faked in order to monitor the results. He even hinted that the 'aerial activity' might be kept secret from other military personnel: 'any hoaxes under a set-up such as this could almost certainly be exposed, perhaps not publicly, but at least to the military.'

Such an experiment, Cross concludes, would help the US Air Force reach clear conclusions about the flying saucer 'problem' and determine how seriously to consider UFO reports, especially those from the general public, should further panics take place. The final line reveals the Air Force's true priorities in dealing with the UFO issue: 'in the future, then, the Air Force should be able to make positive statements, reassuring to the public, and to the effect that everything is well under control'.

Military UFO reports of the period include numerous incidents that could be construed as Stork set-ups, such as during Operation Mainbrace, a huge joint NATO exercise off the northern European coast. In his memoirs, Ruppelt notes that before the exercise began in September 1952, the Pentagon 'half-seriously' told Naval Intelligence that they should keep a lookout for UFOs. There were in fact two significant UFO sightings reported during Mainbrace (one with photographs), both of which sound like large, silver balloons; despite an investigation, however, no department would claim responsibility for them. Had Stork made a special delivery?

The memo from Cross shows that the US Air Force were still struggling to understand what lay at the heart of the flying saucer mystery, and, as other internal memos show, they certainly didn't possess a crashed flying saucer from Roswell, Aztec or anywhere else. So why would the Air Force consider restricting the information that they shared with the CIA? It could be that they simply didn't have any solid conclusions and didn't want the Agency to jump to any of its own; it could have reflected on an inter-service rivalry, such as that between the Air Force and the Navy; or did Battelle and the Air Force want to stage UFO incidents of their own without letting the CIA in on their plans – perhaps hoping to pass the UFO buck on to the CIA? Staging an intrusion into the capital's airspace may seem irresponsible to us now, and it surely was, but it would have sent a powerful message to the heart of the government, and could also have demonstrated that when it came to UFOs the Air Force did, indeed, have everything 'well under control'.

The Washington incident was what finally drove the CIA into taking UFOs seriously; the Robertson Panel was particularly concerned that, as happened the previous July, the Soviets could flood American radars and communication systems with false targets: in a worst-case scenario as a prelude to a nuclear strike. It seems odd, therefore, that no one on the panel mentioned that the technology to do so was already available and may have been responsible for the Washington flap. Was this one of the things that Cross wanted to keep back from the CIA? If so, perhaps the scientist from the unnamed 'agency' who had warned Ruppelt about the upcoming flap was not, as Leon Davidson suspected, from the CIA, but from the Battelle Institute instead?

A clue that at least some in the US Air Force knew what was going on may lie in the fact that General John Samford, the Air Force Intelligence Chief who led the post-flap press conference would, in 1956, become the second-ever director of the National Security Agency. The super-secret NSA monitored international communications and, as we've seen, routinely used the Palladium system in conjunction with the CIA.

Although he didn't know exactly what had been discussed by the Robertson Panel, and he certainly wouldn't have known about the Cross memo, Leon Davidson was convinced that the events of July 1952 were the result of radar-spoofing technology and, rightly or wrongly, had his own ideas as to who was behind the UFO phenomenon. He credits one man with steering the project, CIA director Allen Welsh Dulles, the younger brother of US Secretary of State John Foster Dulles, who ran the Agency like a fiefdom from 1953 to 1961. The first civilian to take the position, Allen Dulles had been deeply involved with wartime intelligence and oversaw the secret importation of German scientists, including V-2 mastermind Wernher von Braun, into the USA under Project Paperclip.

Dulles helped guide the US through the early years of the Cold War; it was under him that the CIA's Directorate of Plans, the 'dirty tricks' department responsible for psychological warfare and covert operations, was formed, just days after the Washington overflights. The Directorate would become critical, and notorious, in developing and maintaining America's role in managing the global status quo. Dulles, as Davidson points out, had also been a close friend and admirer of the philosopher Carl Jung, who wrote presciently about the god-like status of the flying saucer in 1959. The two are sure to have discussed the subject that was so much on everybody's minds during the early 1950s. Davidson was convinced that the CIA were behind each new development of the evolving UFO narrative, with Dulles as its mastermind. Dulles, he wrote, 'co-opted the myth that benign aliens have visited Earth for millennia'[18] and used 'magician's illusions, tricks and showmanship' to transform the usual misidentifications and sightings of military aircraft into alien sightings, landings and contacts.

Why would Dulles and the CIA promote our visitors from space? UFOs provided a handy cover for their own covert psychological and political operations as well as advanced military technology platforms. Spreading the UFO myth might provoke the Russians into wasting time and resources investigating flying saucer stories, among them the possibility that America was flying its own advanced saucer aircraft.

A third reason follows on from Jung's ideas, and also the plot of Bernard Newman's *The Flying Saucer.* The apocalyptic horrors of the Second World War had left many people feeling that God had abandoned humankind and left us to our own vicious devices; that we had entered a new era in which technology had replaced morality as the new religion, and there could be no higher power than the atom bomb. Jung saw in the unbroken, circular form of the flying saucer a modern symbol of god-like perfection: might the belief that there was indeed a higher authority than our own, and that it drove a flying saucer, prevent humanity from ensuring its own self-destruction?

Was Dulles using sophisticated parlour tricks to foment such beliefs? Who, or what, was that higher power? And what did they want to tell us?

SEVEN
PIONEERS OF SPACE

'Unhappily, my friend, Earth is in the hands of raving lunatics; they are unbalanced enough to accuse us of lying if we tell the truth'
Dino Kraspedon, My Contact with Flying Saucers *(1957)*

Looking around the Laughlin Convention at the endless parade of alien statues, figurines, stickers, book covers and blow-up dolls, one might forgive future archaeologists for thinking that twenty-first century humans had worshipped the blank-faced midgets as deities. In reality the relationship between the greys, as the aliens are known, and the people who believe that they've been in contact with them is ambiguous at best. During the height of the alien abduction panic of the 1980s and 1990s, the most common engagement with our alien neighbours came at the wrong end of a scalpel. While most found these close encounters of the surgical kind traumatic, a few determined human guinea pigs were able to transform them into positive experiences, reporting messages of love, peace and impending environmental disaster from their captors. At Laughlin, the aliens seemed to have very little to say for themselves, but it wasn't always this way. Back in the 1950s, following the lead of Klaatu, the preachy humanoid alien from *The Day the Earth Stood Still*, a small missionary force of wisdom-dispensing ETs seemed to descend on planet Earth, telling us how and how not to run our affairs.

Kenneth Arnold's sighting, Maury Island, Roswell and Aztec had marked the dawn of the UFO era, laying down two essential elements of the ongoing mythology: that the UFOs were structured craft and that, like our own planes, they could crash. Until the *LIFE* and the men's magazine *True*, articles most people thought that the craft were American or Russian in origin, but it was now clear that they were from Outer Space – perhaps Venus or Mars, or one of Saturn's moons. The next question, then, was who was flying them?

In 1952, the world found out, thanks to George Adamski, a sixty-one-year-old Polish American who ran the Palomar Gardens Cafe at the base of Palomar Mountain, between San Diego and Los Angeles. Atop the mountain was the 200-inch Hale telescope, then the largest in the world, though Adamski liked to set up his own fifteen and six-inchers for passers-by to peer through at the heavens. Adamski also ran a theosophically inclined mystical order, the Royal Order of Tibet, and lectured regularly at the cafe on esoteric topics to a small coterie of followers who knew him as the 'Professor'.

In 1949 Professor Adamski published a science-fiction novel, *Pioneers of Space: A Trip to the Moon, Mars and Venus*, under his own name, though it was actually written by his secretary, Lucy McGinnis. Soon afterwards he began incorporating flying saucers into his lectures, claiming to have seen and photographed the spacecraft. His reputation as a saucer spotter began to grow locally until, in 1950, he featured in Ray Palmer's *Fate* magazine. Adamski was fast becoming California's one-man flying saucer industry: business at the cafe was booming.

With flying saucers very much in the news, Adamski's mystical saucer group began to attract new members, among them George Hunt Williamson, a close associate of William Dudley Pelley, who ran his own mystical order, the Silver Legion, from Noblesville, Indiana. Pelley, a former Hollywood screenwriter with sixteen films to his name, was an outspoken extremist who hated just about everyone and everything – blacks, Jews, Communists, President Roosevelt – except his hero, Adolf Hitler. Silver Legion chapters existed in almost every state and encouraged their members, known as Silver Shirts, to wear silver Nazi-style uniforms. The Legion also published a number of magazines and, by the early 1940s, had attracted the attention of the FBI. When Pelley began publicly questioning the official story of the attack on Pearl Harbor, he was accused of seditious high treason and sentenced to fifteen years in jail, gaining early parole in 1950.

It was at this time that Williamson and his wife, who had been attempting to contact the flying saucer occupants with an Ouija board, heard Professor Adamski's own tape-recorded communications with the Space Brothers. Deeply impressed, they joined the group and had become key members when, on 20 November 1952, Adamski, Lucy McGinnis, the Williamsons and fellow saucerers Alfred and Betty Bailey saw a large cigar-shaped object drifting over their car as they drove through the California desert. Adamski told his passengers that it was one of the Space Brothers' airships and asked to be left behind.

An hour later, the Professor returned with an incredible story. Alone in the desert with his telescope and camera, Adamski had seen a smaller, beautiful, craft land in the desert about half a mile away. Emerging from the craft was 'a human being from another world', about five feet six inches tall, in his late twenties with long blond hair, high cheekbones and a high forehead. He wore a one-piece brown outfit, red shoes and a perfect smile. Only a few words were spoken, much of the exchange taking place through telepathy and body language.

The Nordic-looking spaceman was called Orthon. He had come to California from his home planet Venus to express his race's concerns for humankind and, in particular, its use of the atom bomb. Orthon asked Adamski to help them spread their message, then, their meeting over, Orthon and his craft took off, leaving behind only the imprint of a single shoe, which Adamski and his friends were able to preserve in plaster of Paris. Strange symbols were visible in the cast, including a star and a swastika.

Adamski was as good as his word and immediately began to talk about his amazing encounter. In 1953 his account (again ghost written, this time by Clara L. John) was included in a bestselling book, *Flying Saucers Have Landed*, along with an essay on flying saucers in antiquity by the Irish peer Desmond Leslie. Adamski travelled the world talking about his meetings with the Space Brothers, which continued after his initial contact. Among those seeking an audience with the Professor were Queen Juliana of the Netherlands and, allegedly, Pope John XXIII. Meanwhile the Space Brothers kept in touch, occasionally dropping by the cafe to catch up with their new Earthling friend, and taking him on trips into Outer Space, which he described in another book, *Inside the Space Ships* (1955).

Later investigations haven't been kind to the Polish professor – his iconic UFO photograph looks uncannily like a chicken feeder and a flying saucer design depicted in a technical paper that was widely circulated in early 1952. What really happened that day in the desert only Adamski, and perhaps Orthon, knew, but the timing of his encounter couldn't have been better, occurring just months after the Washington DC flap and just as the CIA and the US Air Force were discussing how to defuse the saucer issue.

Adamski's stories provided a much-needed answer to the flying saucer question: the occupants weren't evil Russians, they were peace-loving Venusians. While the burgeoning science-minded UFO community, epitomized by Leon Davidson, derided his account, the more spiritually minded tended to accept it, as did the general populace. As Adamski's fame spread, a number of other 'contactees' emerged, all telling similar stories of benevolent Space Brothers and joy rides into Outer Space.

These contactees initiated the first large gatherings of UFO enthusiasts, on a scale that would put the Laughlin Convention to shame. At conventions like those held for many years beneath Giant

Rock in the Mojave Desert, organized by contactee and Douglas aircraft engineer George Van Tassel, the UFO faithful in their thousands could hear and share all the latest news and theories on their favourite subject. The contactee wave set a new tone for public and government attitudes towards the UFO phenomenon and presented an image of the UFO enthusiast starkly at odds with the serious approach of Davidson or Keyhoe. Gone was the earnest scientist seeking answers to a technological riddle; in his place were kooks, mediums and crazies.

Speculation that Adamski and some of the other contactees were embroiled in intelligence games has existed since the 1950s. Leon Davidson contacted Adamski shortly after he went public with his encounter, exchanging several letters with him over the years. Asked by Davidson if there was anything at all unusual about Orthon and his fellow Space Brothers, Adamski replied that he was 'very definitely a human being . . . with his hair cut and in a business suit as men here wear he could mingle with anyone, anywhere'.[1]

Davidson naturally sensed Allen Dulles's stagecraft behind Adamski's encounters. His later 'trips' to Outer Space, documented in *Inside the Space Ships*, always began with the Space Brothers picking him up in a black Pontiac and driving him out into the desert. Here Adamski would enter a landed 'scout craft', inside which he would sit in a chair and watch the stars zoom by on a pair of screens (the craft's portholes were always closed), though he sensed 'no motion at all' while flying. During these trips 'newsreels' from Venus would be shown on the screens and the Space Brothers would lecture on various topics, all the while serving Adamski oddly coloured drinks. A suspicious Davidson noted that in 1955 a 'Rocket to the Moon' ride had opened at Disneyland that used back-projections to create the sensation of spaceflight. Were Adamski's space journeys concocted using similar Hollywood special effects? And just what was in those in-flight drinks that the Space Brothers were serving up to him?

Whoever was behind Adamski's adventures, the involvement of Silver Shirt George Hunt Williamson in his circle, combined with persistent rumours that the Royal Order of Tibet masked a moonshine operation during Prohibition, would have been enough on their own to warrant the FBI's undivided attention. And Adamski had it, from an early stage in his career. A September 1950 FBI report paints a vivid picture. 'If you ask me,' the Professor told an FBI agent, 'they probably have a Communist form of government . . . That is a thing of the future – more advanced.' He also predicted that 'Russia will dominate the world and we will then have an era of peace for 1,000 years.' He pointed out that Russia already had the atom bomb and that:

> within the next twelve months, San Diego will be bombed . . . The United States today is in the same state of deterioration as was the Roman Empire prior to its collapse and it will fall just as the Roman Empire did. The Government in this country is a corrupt form of government and capitalists are enslaving the poor.'[2]

Not all the contactees were Communists: for example Carl Jung's favourite, Orfeo Angelucci who, like George Van Tassel, also worked in the aerospace industry, was clearly on America's side: 'Communism, Earth's present fundamental enemy, masks beneath its banner the spearhead of the united force of evil . . . [it] is a necessary evil and now exists upon Earth as do venomous creatures, famines, blights, cataclysms – all are negative forces of good in man and cause them to act.'[3]

While there is clear evidence of FBI monitoring of the contactees, whether or not some of them were actively encouraged or manipulated by American, or even Soviet, government agencies we don't know. Leon Davidson had no doubt that the CIA were 'working' Adamski and his fellow travellers, and saw their predominantly peaceful, anti-atom-bomb message as a key factor in the growth of the international peace movement, which culminated, in 1958, with a short-lived ban on nuclear tests agreed between the US, the UK and the Soviet Union. Was this all part of Allen Dulles's master plan?

Whether they were CIA puppets or not, Adamski and the other contactees put into action some of the key recommendations of the Robertson Panel report: they ensured that it would be some time before the UFO subject would again be taken seriously, while their conventions herded the UFO faithful together, making keeping tabs on them all that much easier for the intelligence agencies, a tradition that continues to this day.

MAURY ISLAND, BRAZILIAN STYLE

For Adamski and his fellow contactees, encounters with the space people were profound experiences and a ride in one of their vehicles was a thrill like no other. The Space Brothers exuded a benevolence and wisdom befitting their advanced intellectual and technological means, a wisdom that their human passengers were inspired to share with others. But what if, instead of space preachers, you met alien fiends who kidnapped, drugged and raped you, all the while grunting and barking like overexcited dogs? This is exactly what happened to a young Brazilian farmer, Antonio Villas Boas in 1957.

A curious prelude to the Villas Boas story begins in Rio de Janeiro in September of that year. In his popular column in Rio's *O Globo* newspaper, Ibrahim Sued reported that he had been sent a fragment of a flying saucer by one of his readers. Sued, whose column usually dealt with celebrity gossip, had never taken an interest in UFOs before, but he reprinted part of the letter:

> As a faithful reader of your column and your admirer, I wish to give you something of the highest interest to a newspaperman, about the flying discs . . . just a few days ago . . . I was fishing together with some friends, at a place close to the town of Ubatuba, São Paulo, when I sighted a flying disc. It approached the beach at unbelievable

> speed and an accident, i.e. a crash into the sea, seemed imminent. At the last moment, however, when it was almost striking the waters, it made a sharp turn upward and climbed rapidly on a fantastic impulse. We followed the spectacle with our eyes, startled, when we saw the disc explode in flames. It disintegrated into thousands of fiery fragments, which fell sparkling with magnificent brightness . . . Most of these fragments, almost all, fell into the sea. But a number of small pieces fell close to the beach and we picked up a large amount of this material which was as light as paper. I am enclosing a small sample of it.[4]

The column was read by Dr Olavo Fontes, a respected young medical doctor who would become Vice President of the Brazilian Society of Gastroenterology and Nutrition before dying of cancer in 1968, while still in his thirties. Fontes had become fascinated by dramatic reports of UFOs over Brazil in late 1954 and, after starting to investigate individual cases on his own, joined APRO, the American UFO organization recommended for observation by the Robertson Panel, in early 1957.

After reading about the crash in Ubatuba, Fontes immediately contacted Sued and arranged to have the light, metallic material analysed at the National Department of Mineral Production in Brazil's Agricultural Ministry. Samples were also sent to the US Air Force, via the US embassy. The material turned out to be magnesium, some samples of which were unusually pure, while others had odd compositional elements, though these could easily have been added to it.[5]

The investigation made the national news, launching Fontes's career as Brazil's foremost ufologist and setting him up for an odd close encounter of his own. In February 1958 Fontes was visited by two Brazilian Naval Ministry intelligence officers who wanted to talk to him about the Ubatuba material. After warning him not to poke his nose into matters that 'did not concern him', they proceeded to tell him everything they knew about the secret UFO cover-up. The world's governments, they said, were aware of the extraterrestrial presence on Earth and were doing everything they could to keep a lid on it. Six flying saucers of between thirty and a hundred feet in diameter had crashed thus far, three in the US (two in good condition) one in the UK, one in the Sahara and one in Scandinavia.[6] All of the craft had contained short, humanoid occupants, none of whom had survived, and scientists were currently trying, unsuccessfully, to back-engineer these saucers, which seemed to be powered by strong rotating electromagnetic fields, along with an atomic component. The UFO occupants themselves had shown no interest in contacting humankind and were to be considered extremely hostile, having already destroyed a number of aircraft sent to pursue them. The UFO matter, they warned him, was held at the highest level of secrecy – even Brazil's president was kept in the dark on the matter – and was considered sensitive enough that some witnesses and researchers had been assassinated to prevent them from leaking information.

Fontes was left puzzled but unbowed by the visit. He may even have asked the same questions that we should: why, if the UFO matter was so secret that even the president couldn't be told, had so many of the Navy men's revelations already been printed in popular books and magazines? And why were they telling Fontes, who immediately shared the information with Coral and Jim Lorenzen APRO's directors, confirming similar rumours that they had heard form other sources? Was it because somebody *wanted* Fontes and APRO to believe these tales, and to share them, in the same way that Silas Newton had been encouraged to keep spreading his crashed saucer stories back in 1950?

ABDUCTEE ZERO

The timing of Fontes's visit by *os homens de preto* was serendipitous, if not sinister, coming just days after the doctor had first met a young farmer with a strange and terrifying tale of kidnappers from Outer Space.

It was 16 October 1957, less than two weeks after Sputnik had become the first man-made object to orbit the Earth, when they took Antonio Villas Boas. The twenty-three-year old was ploughing his family's fields near São Francisco de Sales, on the banks of the Rio Grande in Brazil's south-eastern Minas Geris state. He was working alone and at night to avoid the heat of the sun, and this made him uneasy.

Two nights previously he and his brother João had been ploughing the same fields when a glaring red light had startled them. It was bright enough to hurt their eyes, and occasionally gave out dazzling blasts 'like the setting sun'. Every time they tried to approach the light it would zip away until, suddenly, it vanished as if 'it had been turned out'.

At 1 a.m. on 16 October, the red light returned, 'lighting up the tractor and all the ground around, as though it were daylight'. Moments later the object landed about 150 feet from the bank of the Rio Grande. As it did so, the tractor's petrol engine died and its lights went out.

'It was a strange machine,' Villas Boas told Fontes. 'Rather rounded in shape, and surrounded by little purplish lights, and with an enormous red headlight in front . . . [it was] like a large, elongated egg . . . On the upper part of the machine there was something which was revolving at great speed and also giving a powerful fluorescent, reddish light.'[7] The following day, Villas Boas measured the tripod mark left by the craft and estimated it to be thirty-five feet long and twenty-three feet across at its widest point.

When the flying machine landed Villas Boas tried to run, but a short, strong individual dressed in 'strange clothing' grabbed him violently. Three taller beings then emerged and carried him back to their craft, pushing him up a metal ladder, then through a hatch with a drop-down door. Villas Boas provided a detailed description of the beings' clothing. They wore grey one-piece overalls decorated with black stripes; on their heads were cloth helmets reinforced with thin metal strips, including a triangular one on the nose between two lensed eyeholes. The tops of the helmets extended almost twice the height of a normal human head, giving the impression of large foreheads (a detail shared by Adamski's Orthon), and from these emerged thin silvery tubes that ran into the backs of the overalls. The abductors wore stiff, five-fingered gloves, rubber thick-soled boots and a single, round red reflector on their chests, the size of a pineapple slice.

It all sounds more than a little low-fi, like something from a Buck Rogers serial or, indeed, *The Day the Earth Stood Still*. The interior of the craft also seems to reflect a rather 1950s vision of the future. It was round, white, brightly lit and featureless, the only furniture being a metal table and swivelling stools, all bolted to the floor. A ring of square fluorescent lights ran around the ceiling.

Communicating with guttural barks and yelps, the humanoids stripped their captive, wiped him down with a wet sponge and took him to a room containing a large, soft bed covered by a grey sheet. Here they placed a device like a cupping glass that draws blood under his chin, then he was left alone. A grey smoke filled the room, entering from holes in the walls, which left him feeling nauseous as a short, extremely beautiful and entirely naked woman appeared in the doorway. She was human, though her features were unusually pointed. Her hair was almost white and parted in the centre, though her pubic hair was bright red; her eyes 'were large and blue, more elongated than round, being slanted outwards like the slit eyes of those girls who make themselves up fancifully to look like Arabian princesses'.[8]

The woman began to rub herself against Villas Boas, leaving him uncontrollably aroused. One thing led to another: 'It was a normal act,' said Villas Boas, 'and she behaved just as any woman would.' The farmer blamed his excitement on the fluid that his captors had wiped over his body, though one wonders if an aphrodisiac would have been necessary under the circumstances. When they were finished the woman smiled, pointed at her stomach and then to the sky, leading Villas Boas to conclude that she was going to have their hybrid child.

Once dressed he was shown around the craft's exterior, then told that it was time to leave. The perplexed and shaken farmer watched as the craft took off with a loud whining sound; its lights flashed different colours before settling on bright red and the rotating top spun ever faster as it began to lift up slowly from the ground, its three legs telescoping up into its belly. It rose up a hundred feet, buzzing loudly, then, with a sudden jolt, shot into the air like a bullet. The young farmer was deeply impressed. 'These people really knew their business,' he told Dr Fontes, stating his belief that they were humans, though perhaps from another planet.

It was now 5.30 a.m.; the incident had lasted about four hours. When Villas Boas tried to start his tractor he found that it was still dead. The engine's battery wires had been unscrewed – a low-tech but effective means of preventing his escape. He staggered back home, where his sister recalled that he vomited a yellow liquid and had dark bruises on his chin. For the next few weeks he suffered aches and pains, eye irritation and multiple small lesions.

Soon after this incident he wrote to João Martins, editor of the popular *O Cruzeiro* magazine, who flew him to Rio de Janeiro, where he was interviewed and examined by Dr Fontes. Martins never printed Villas Boas's story, which instead appeared in an obscure Brazilian UFO journal in 1962, only reaching the English-speaking UFO press in the mid-1960s. Fontes was impressed by the sincerity of the young farmer and was convinced by his account, despite its outlandish nature. Villas Boas never sold his story to the newspapers even though Martins suggested that he could profit from doing so.

So what happened? Was Antonio Villas Boas really kidnapped by sex-starved aliens? Did he dream or hallucinate the entire account – perhaps after knocking himself out, explaining the bruising on his chin? It's certainly within the bounds of normal human psychology: he and his brother had seen the red light in the sky earlier that month, and there had been reports of UFO sightings in the papers; perhaps these inspired his alien fantasy? There is another possibility, however, one that might seem as absurd as a real alien kidnapping.

THE BRAINWASHING MACHINE

Until his death in Fairfax, Virginia in 1999, Bosco Nedelcovic was an interpreter and translator at the Inter-American Defense College, which educates future leaders of Latin American nations. In 1978 the Yugoslavian émigré confided in the American UFO researcher Rich Reynolds that, during the 1950s and 1960s, the CIA had deliberately manufactured UFO incidents all over the world as part of a project called Operation Mirage.[9] What's more, Nedelcovic, who between 1956 and 1963 had worked for the CIA in Latin America under the Agency for International Development (AID), was himself present at some of these staged events. And one of them was the Villas Boas abduction.

Nedelcovic claimed that in mid-October 1957 he was part of a helicopter team conducting psychological warfare and hallucinogenic drug tests in the Minas Gerais region of Brazil. The team included himself, two other CIA employees, a doctor, two naval officers – one American and one Brazilian – plus a three-man crew.[10] Various items of electronic equipment were on board, as was a metal 'cubicle' about five feet long and three feet wide. Nedelcovic was never told what it was for, only that it was used by the military in psychological warfare operations.

Initially the team flew around their base of operations at Uberaba, about 150 miles east of São Francisco de Sales, testing the electronic equipment. A few days later they flew along the Rio Grande and conducted another night sweep. Using heat-sensing cameras they identified a lone figure on the ground. The helicopter descended to about 200 feet then released an aerosol sedative.[11] The helicopter landed, and the man ran, pursued by the three CIA operatives who grabbed him and hauled him into the helicopter, banging his chin on the deck as they did so. Nedelcovic makes no mention of what was done to the man while on board, only that after a few hours they left him, still unconscious, next to his tractor.

So was this man Antonio Villas Boas? Elements of Nedelcovic's account do correspond to the farmer's story, for example the timing, geography, weather conditions and bruising under their victim's chin. Likewise many aspects of Villas Boas's account, his kidnappers' costumes for instance, sound more human than alien. Their aircraft also sounds as if it could have been a helicopter, modified either by his own imagination or by some crafty sci-fi refitting; the unusual number of coloured and white lights on the exterior may have given it an extra UFO-like feel, while the 'rotating' dome on its top might have been rotor blades. Countering this, a large helicopter makes a lot of noise, and it would be some years before 'quiet' helicopters were operational in the field. Even late at night in such a remote area, somebody would surely have heard the chopper, an unusual sound in rural Brazil at the time.

Other parts of the story also ring true. At the time of the incident the CIA and the US military were firmly established in Brazil, and all over Latin America, keeping close tabs on political developments in the region. Brazil was considered a particularly sensitive nation; its vast size, considerable natural resources and proximity to the US made it a ripe target for Soviet expansion. Things would come to a head in 1964 when the CIA participated in a coup to oust President João Goulart, replacing him with a brutal military junta that held power for the next two decades.

In 1957 the CIA was also deeply involved in its MK-ULTRA programme, researching mind and behaviour-altering techniques involving drugs, surgery and technology. They experimented with a number of psychoactive substances – hallucinogens, sedatives, stimulants, psychomimetics and more – often on entirely unwitting subjects. Would the CIA have conducted some tests on subjects outside of their own jurisdiction? Certainly. For the CIA at this time, the entire world was within its jurisdiction.

Villas Boas was repeatedly sick during and after his experience, and also suffered unpleasant physiological effects that Fontes took to be related to radiation exposure. Might the strange 'cubicle' described by Nedelcovic have been used to illicitly test the effects of radiation exposure? Although Nedelcovic doesn't mention what they wore for the helicopter flights, Villas Boas's description of the entities' clothing and helmets could be conceived of as radiation-protection gear.

Following this line of enquiry just a little further, we might ask whether it's possible, under hypnosis or the influence of hallucinogenic drugs, to make somebody believe that they've experienced something that they haven't. The answer is a firm yes – and it doesn't take much. In 2001 psychologists at the University of Washington showed people who had been to Disneyland as children mock advertisements featuring Bugs Bunny at the theme park.[12] Sometimes a large cardboard cutout Bugs was placed in the room with the subjects. When questioned later, a third of the group who saw the advert remembered meeting Bugs Bunny at Disneyland, as did 40 per cent of those who viewed the ad with the cutout in the room. The subjects were neither hypnotized nor on drugs, but they could not have met the giant talking rabbit at Disneyland – Warner Bros. and Walt Disney's lawyers would never have let it happen.

Creating false memories is relatively easy, but when it comes to UFO entity encounters, this is a knife that cuts both ways. In some individuals, simply reading about UFO incidents, or alien abductions, combined with unusual though not uncommon experiences such as sleep paralysis or fugue states, might lead them to suspect that they had undergone a real-life encounter of some kind. This may be what happened to Villas Boas; his experience could be a vividly remembered fantasy with no basis in actuality. But if we combine Elizabeth Loftus's research with Nedelcovic's claims, a different picture emerges.

In his 1959 book *The Brainwashing Machine*, the Hungarian author Lajos Ruff describes being snatched by the Communists in 1953 and placed in his prison's 'Magic Room', every aspect of which is designed to psychologically destabilize a drugged subject. The walls are round; furniture is bolted to the floor; strange, psychedelic lighting effects are deployed, including rotating coloured gels and a focused laser-like beam; photographs and films of sex and violence are flashed on to a screen. On one occasion Ruff was shown a film of a woman having sex with a man whose face was obscured. He awoke later to find the same woman lying next to him, talking about the film as if it had really happened; she then had sex with him. In the Magic Room, reality and fantasy were blurred in an attempt to psychologically 'break' the victim.

Ruff escaped to the US and testified to Congress about his experiences in a hearing about the practice of 'brainwashing'.[13] But his story is not that simple. While Ruff probably did endure awful psychological tortures while imprisoned in Hungary, as a political asylum seeker in America he would have been encouraged to dramatize his experience both for Congress and the public. His lurid and sensational book *The Brainwashing Machine* would have been a part of that process and may have been written by his CIA handlers – a common propaganda practice during the Cold War.

Even if Ruff's Magic Room was a fiction, however, MK-ULTRA was not. Could Villas Boas have been the victim of CIA psychological manipulation inspired by pulp science fiction and the UFO mythology developing in America since 1947? We would have to say yes; maybe.[14]

EIGHT

AMONG THE UFOLOGISTS

'I found myself in a circular, domed room . . . made of an ethereal mother-of-pearl stuff, iridescent with exquisite colors that gave off light. Then I heard music! I recognized the melody as my favorite song, "Fools Rush In". I realized how safe I was with them – they who knew my every thought, dream and cherished hope!'

Orfeo Angelucci, The Secret of the Saucers *(1955)*

From my hotel room at the Laughlin Flamingo, I looked out at the array of flashing, blinking neon lights whose glow darkened the stars in the desert night. I may not have been any closer to finding out what my friends and I had seen that sunny day in 1995, but I did have a clearer sense of how the stories that I, John and so many millions of others were fascinated by had begun to take shape.

Believing that the military and the intelligence agencies were behind the entire flying saucer phenomenon struck me as being no less misguided or paranoid than any of the other wild tales circulating within the UFO lore. Yet it seemed clear that the US Air Force, the Navy, the CIA, the NSA and who knows which other members of this cryptic alphabet soup had knowingly deceived the public and, at times, each other, about UFOs. Each had, in their own way, exploited the phenomenon to their own ends and, in doing so, shaped the way that the mythology had unravelled. Whether the UFOs were flying overhead, crashing to the ground, hailing us or kidnapping us, there were human fingerprints all over them.

But I was frustrated. So far this strange story had been pieced together from old books and articles, the occasional government document, smatterings of anecdote and hearsay and a considerable amount of speculative intuition. There was little in the way of hard evidence, but what could I expect? Secrecy is the very air that the intelligence services breathe, and if clandestine UFO operations were still going on then their history was hardly likely to be an open book. 'Estimating' – an old intelligence saying goes – 'is what you do when you don't know and cannot find out.'

The strategists, operators and agents behind the first wave of UFO lore, the original Mirage Men, were all long gone now, and John and I would never get to talk to them. But by following Bill Ryan into the Serpo story, and with the Laughlin Convention now in full swing, we hoped that we would soon get to meet their progeny.

FULL DISCLOSURE

In 1953 the CIA Robertson Panel had recommended that civilian UFO organizations should be closely monitored (for 'monitored' we can probably read infiltrated), mentioning the Aerial Phenomena Research Organization (APRO) and Civilian Saucer Investigations (CSI) by name. If the wiser members of the UFO community were aware that they were being watched and sometimes interfered with by the government, they tended to believe that it was because they were getting too close to the truth of extraterrestrial visitation. Three decades later, a very different picture of government involvement began to emerge, one that most ufologists, perhaps understandably, chose to ignore. It all hinged on ufology's first whistleblower, a heroic researcher turned traitor and pariah: enter William Moore.

Bill Moore was one of the most respected players in the field. He'd been largely responsible for digging up the Roswell story after forty years of obscurity, and his bestselling book *The Roswell Incident* had contributed to the field's increasingly presentable public image. But by the time of his presentation at the 1989 Mutual UFO Network (MUFON) Conference at the Aladdin casino hotel in Las Vegas, the UFO community was in total disarray: the conference reflected what was, essentially, a civil war. As the relatively sober-minded official MUFON event took place at the Aladdin, a splinter conference was being held nearby at another site. The speakers here advocated the more extreme, 'dark side' of the UFO phenomenon, warning of the successful alien colonization of the planet and a vast government conspiracy to cover it up while providing human genetic material to the ETs, harvested in terrifying abductions, in exchange for advanced military technologies.

As Moore took to the stage at the 'official' conference, he knew that this could be his last public appearance. He knew this because he was in no small part responsible for the chaos that had engulfed ufology. And he was about to tell the world.

A grainy videotape is all that remains of Moore's speech; the sound, recorded separately, drifts in and out of sync, occasionally dropping out entirely. Moore, a bear-like man in a what might be a grey or brown suit, his face mostly hidden by a thick beard, dark glasses and a bowl haircut, looks nervous behind the podium, shifting on his feet before the standing-room-only crowd of about a thousand people. He coughs and begins: 'Ladies and gentlemen, friends and adversaries, associates and colleagues, in short, fellow ufologists. Allow me to introduce myself. I'm aware you already know my name, and I'm equally certain that many of you are wondering just who is Bill Moore, and what's he up to? . . . I have known exactly what I was doing all along, it was just that nobody else knew what I was doing . . . For some of you it may not be what you want to hear, but it is, nonetheless, true.'

Over the next two hours Moore, firm but restrained, told his story[1] – and proceeded to tear the UFO community a new hole from which to view the skies. Following the success of two book collaborations with Charles Berlitz (himself a former Army Intelligence officer) – *The Philadelphia Experiment* (1979) and *The Roswell Incident* (1980) – Moore abandoned his job as a labour-relations expert and moved to Arizona to write full time. He also became Head of Special Investigations for APRO, who were based in Tucson.

In early September of 1980, following a radio appearance to promote *The Roswell Incident*, Moore had received a telephone call at the station. 'We think you're the only one we've heard that seems to know what he's talking about,' said the anonymous caller, with a hint of an Eastern European accent; then he hung up. A few days later Moore was at another radio station, in Albuquerque, New Mexico. The man called again, repeating his message. This time Moore agreed to meet him at a local restaurant; the caller would be wearing a red tie. Moore named his contact Falcon, after Robert Lindsay's book *The Falcon and the Snowman: A True Story of Friendship and Espionage*. He has never revealed Falcon's position or his identity, stating only that he was 'well placed in the intelligence community'.

Over dinner Falcon offered Moore a deal. If Moore played his cards right, Falcon could offer him what the UFO community sought most of all: incontrovertible proof of a government UFO cover-up. In exchange, Moore was to provide Falcon with what the intelligence community sought most of all: information. 'I knew I was being recruited,' said Moore in 1989, 'but I had no idea for what.' Moore agreed to the proposition and was handed a manila envelope. Inside was an Air Force document: it referred to 'Project Silver Sky' and described civilian sightings of an aerial 'object' and the recovery of a 'spike craft'.

Moore and Falcon met again on 30 September. Accompanying Falcon was a young Air Force Office of Special Investigations (AFOSI) agent, Richard Doty, who was to be Moore's conduit for communicating with Falcon. Moore immediately took the pair to task over the Silver Sky document. He had investigated the names of the witnesses and none of them checked out: the document was a forgery. Falcon congratulated him on passing their first test. He was now ready to take the next step.

Falcon revealed that he worked for the Defense Intelligence Agency (DIA) and represented a group within the intelligence community who wanted to tell the public the truth about UFOs. Moore was someone they could trust to help them do it. In return, Moore would be asked to pass them information about who was doing what within the UFO community, and to reverse the flow by feeding false information back into the UFO field. Moore was being inducted into the disinformation game.

In fact the game had already begun. Earlier that year Moore had been handed a strange letter by the directors of APRO. Written by Craig Weitzel, a young Air Force cadet based at Kirtland AFB in Albuquerque, New Mexico, it described a landed UFO and a silver-suited occupant seen by ten cadets while on a training exercise. Weitzel had photographed both the craft and the being. Back at Kirtland, Weitzel had been approached by a mysterious dark-haired man in a dark suit and shades. His name was Huck, he said; he was from Sandia Labs on the base and he wanted the UFO photographs. Spooked, Weitzel handed them over. The cadet wrote that he had reported the incident to a Kirtland security officer named Dody, and ended the letter with the statement that a crashed UFO was being stored beneath the Manzano mountains on the base.

Ever the vigilant investigator, Moore had tracked down Weitzel, who told him that he really had seen a silvery UFO that had accelerated away suddenly 'like you never saw anything accelerate before'. But it was not at the site reported in the letter, there had been no occupant, he had taken no photographs and he certainly hadn't received a visit from the sinister Mr Huck. Mr Dody, of course, was Richard Doty, and the Weitzel letter was bait. In Bill Moore, AFOSI had landed the catch they were fishing for.

It's unclear whether it was APRO or AFOSI who first suggested that Moore drop in on Paul Bennewitz. Both had their reasons. Bennewitz had begun filming strange lights zipping over the Manzano mountains in July 1979. He was also picking up strange signals on his home-made receivers that he felt sure were coming from these UFOs. Things took a turn for the weird in May 1980, when Bennewitz was introduced to Myrna Hansen, a young mother. Hansen and her eight-year-old son had seen a strange bright-blue light over her car while driving near Eagle's Nest, about sixty-five miles north-west of Albuquerque. She had reported it to APRO, who suggested that she visit Bennewitz, their representative in the area. Under hypnosis with Bennewitz, Hansen had described being lifted into a UFO, where she saw terrible things, including mutilated cattle and human body parts stored in vats. Afraid of what had happened and trusting the man who seemed to have the answers, Hansen became a regular visitor to the Bennewitz home and the pair slipped into a disturbing *folie-à-deux*.

Hansen had put her trust in Bennewitz as a scientist and a UFO expert, but his letters to APRO were becoming increasingly bizarre. He announced, for example, that Hansen's alien abductors had implanted a tracking device in her body that allowed them to follow her every move and control her every thought. Concerned for Hansen's safety and Bennewitz's state of mind, APRO asked Bill Moore to go and check him out. The US Air Force, meanwhile, also wanted Moore to pay Bennewitz a visit, to gauge how well the disinformation they were already feeding him was working out.

Internal documents from Kirtland Air Force Base, released via the US Freedom of Information Act, reveal the alarming speed with which the campaign against Bennewitz had swung into action. The physicist first reported his sightings to Kirtland's head of security, Major Ernest Edwards, on 24 October 1980. Edwards then passed on the matter to Doty at AFOSI. Their report describes what happened next.

On 26 Oct 1980, [Special Agent] Doty, with the assistance of Jerry Miller[2] [a] Scientific Advisor for Air Force Test and Evaluation Center . . . interviewed Dr. Bennewitz at his home in the Four Hills section of Albuquerque, which is adjacent to the northern boundary of Manzano Base. Dr. Bennewitz . . . produced several electronic recording tapes, allegedly showing high periods of electrical magnetism being emitted from Manzano/Coyote Canyon area. Dr. Bennewitz also produced several photographs of flying objects taken over the general Albuquerque area. He has several pieces of electronic surveillance equipment pointed at Manzano and is attempting to record high frequency electrical beam pulses. Dr. Bennewitz claims these Aerial Objects produce these pulses.

. . . After analysing the data collected by Dr. Bennewitz, Mr. Miller related the evidence clearly shows that some type of unidentified aerial objects were caught on film; however, no conclusions could be made whether these objects pose a threat to Manzano/Coyote Canyon areas. Mr. Miller felt the electronical [sic] recording tapes were inconclusive and could have been gathered from several conventional sources. No sightings, other than these, have been reported in the area.'[3]

On 10 November, Bennewitz was invited to Kirtland to present his findings to the heads of the various units and departments on the base. By the end of his presentation the only people remaining were representatives from AFOSI and the NSA, who had realized that Bennewitz was somehow picking up their own experimental transmissions that were, as far as we know, unconnected to the lights that he was filming. The NSA decided to let him continue intercepting the signals – that way they could work out how he was doing it, and whether or not he could be of use to them.

On 17 November, AFOSI's new recruit, Bill Moore, was summoned to their offices for the first time and asked to tell the team what he knew about the state of UFO research. Following the meeting, Richard Doty showed Moore a teletype display for classified internal AFOSI communications on which a new document had appeared. Labelled 'Secret', it contained an analysis of three photographs and two strips of 8mm film shot by Bennewitz. At the bottom of the document was a reference to Project Aquarius.

In February 1981 Falcon and Doty asked Moore to hand a document to Bennewitz. At first glance it appeared to be the same one that he had seen on the teletype in November, but on closer inspection he noticed that it had been subtly altered. The document read:

> (S/WINTEL) USAF no longer publicly active in UFO research, however USAF still has interest in all UFO sightings over USAF installation/test ranges. Several other government agencies, led by NASA, actively investigates [sic] legitimate sightings through covert cover. (S/WINTEL/FSA)
>
> One such cover is UFO Reporting Center, US Coast and Geodetic Survey, Rockville, MD 20852. NASA filters results of sightings to appropriate military departments with interest in that particular sighting. The official US Government Policy and results of Project Aquarius is still classified top secret with no dissemination outside official intelligence channels and with restricted access to 'MJ Twelve'. Case on Bennewitz is being monitored by NASA, INS, who request all future evidence be forwarded to them thru AFOSI, IVOE.'[4]

In 1983 Air Force Intelligence conducted an investigation into this document under the heading 'Possible Unauthorized Release of Classified Material'. They noted several problems, not least among them that its alleged source, a Captain Grace, didn't exist and that the format was incorrect for classified documents. The rather irritating report also complained that 'the document is replete with grammatical errors, typing errors, and in general, makes no sense . . .'[5] In other words, it was a fake, and not a very convincing one to those who knew what they were looking for. Beyond the technicalities there were more fundamental problems with the document. The idea that NASA would be monitoring Bennewitz was absurd: they had no facilities at Kirtland, and were not in the surveillance business; the US Coast and Geodetic Survey, meanwhile, had ceased operations in 1970.[6] The document also made reference to 'MJ Twelve', an organization that nobody at the time had heard of; it was a name that would haunt the UFO community for the next thirty years.

Falcon and Doty were insistent that Moore pass the Project Aquarius document to Bennewitz. This was the first time they had asked him actively to deceive somebody else and initially Moore refused to cross the line: he and Bennewitz were now in regular contact and were becoming friends. But his handlers made it clear that if he wouldn't play the game, then their relationship would end then and there.

That summer, Moore handed the document to Bennewitz at his Thunder Scientific lab. This was the proof Bennewitz had been waiting for: it confirmed that the Air Force was taking his research seriously and that, therefore, he was on the right track. In a cupboard room, with the radio turned up to mask their conversation – Moore knew that Thunder Scientific was being bugged – he pleaded with Bennewitz not to share the document with anybody else, but it was to no avail. The sightings and signals over Kirtland, combined with the information extracted from Myrna Hansen, had convinced Bennewitz that an alien invasion was taking place on his doorstep. He had told the US Air Force, but now he needed to warn the world.

Bennewitz wrote letters to APRO, to local politicians including the Senator of New Mexico, to former astronaut Harrison Schmitt, and even to President Ronald Reagan. This last letter earned him a reply from the office of the Air Force secretary, reeling out the standard line that the Air Force had stopped investigating UFOs with the closure of Project Blue Book in 1969. Bennewitz knew that this was untrue: he could name a handful of Air Force personnel who were investigating UFOs right at that moment at Kirtland.

Cranking up his campaign against the invaders, Bennewitz modified his computer system to decode the 'alien' transmissions, translating repeated signals into specific words such as 'UFO', 'spaceship' and 'abduction'. Bennewitz felt that if he could understand the aliens' plans, he could help the Air Force to fend off their invasion, so he began passing the decoded messages to his friends at AFOSI, who continued to take great interest in what he was doing.

In mid-1981 Bennewitz received a distinguished visitor, Professor J. Allen Hynek, the famed astronomer and UFO researcher, who had brought a new computer with him. This was a visit from UFO royalty and, once again, would only have confirmed to Bennewitz that he was on to something important. Hynek had been closely involved with UFOs since he was hired to assist the US Air Force's Project Sign in 1948. Initially dubious of the subject, Hynek became convinced that the phenomenon represented something real and unknown, and in the late 1960s he turned against the Air Force. In his 1972 book *The UFO Experience* Hynek denounced Project Blue Book as a sham and introduced his own system for categorizing UFO encounters, which included 'Close Encounters of the Third Kind' to describe encounters with beings on board a UFO. The term gave Steven Spielberg the title for his 1977 film, which features a cameo from Hynek, puffing his trademark pipe, during its spectacular conclusion.

At the time of his visit to Bennewitz, Hynek was running the Center for UFO Studies while holding a professorship at Evanston University and, allegedly, receiving a $5,000-a-year honorarium from the US Air Force. Would Hynek have knowingly taken part in the Air Force campaign against Paul Bennewitz? It would certainly shoot holes in his image as the Colonel Sanders-like kindly 'godfather' of scientific ufology. But Bill Moore claims that Hynek himself told him that the Air Force instructed him to deliver the computer, which contained their own special software, to Bennewitz, and not to tell him where it came from.

Bennewitz set up the new computer and began feasting on the vastly 'improved' decoded messages that it delivered to him. A sample of these new texts reveals how consumed Bennewitz was by his alien fantasy: 'VICTORY OUR BASES OBTAIN SUPPLIES FROM THE STARSHIP METAL TIME IS YANKED TIME IS YANKED MESSAGE HIT STAR USING REJUVINATION [sic] METHODS GOT US IN TROUBLE SIX SKYWE REALIZE TELLING YOU ALL MIGHT HELP YOU.'[7] There are pages and pages of this material. Moore thought that the new computer assigned 'entire words, sentence fragments, or sometimes even entire sentences themselves to the various individual pulses of energy'[8] being transmitted into Bennewitz's home. Somebody was fuelling the flames of Paul Bennewitz's fantasy. But who?

As Bennewitz's letters about his 'research project' drew more attention to Kirtland, so the surveillance increased. He became convinced that his house and car were being broken into and that the house opposite his was occupied by agents. Moore remembered a white van pulling up alongside the two men one day, long enough for a man to photograph them. He traced the licence plate to North American Aerospace Defense Command in Colorado: it seemed that Four Hills, Albuquerque, had become a playground for the nation's intelligence services.

None of this was doing Bennewitz's mental health much good. Once, in 1982, Paul pointed out lights over a Mesa region on the far side of the base, telling Moore that they were UFOs abducting humans for genetic experiments. When Moore went to investigate, he found military helicopters flying over the area with searchlights on a search and rescue training mission, a fact confirmed by Kirtland. But it wasn't all fantasy. On one visit to the Bennewitz household, Moore saw a ball of yellow light, about twelve centimetres across, hovering near the ceiling of Paul's home lab. The light wobbled slightly and was surrounded by a pale-blue glow.

When Moore expressed his surprise, Bennewitz said that the orbs kept showing up but he couldn't work out what they were.

Interviewed by Greg Bishop twenty years later, Richard Doty claims that he and two National Security Agency (NSA) agents observed one of these balls as they were snooping around the house one night. 'It was orange with sparkles in it,' Doty remembered. Doty asked the NSA men if it belonged to them – they were set up over the road at this time, beaming signals at the Bennewitz home – but they denied it. All three tried to see if the light was being projected from anywhere, but were unable to find its source.

Within a year of making contact with the security forces at Kirtland, Bennewitz had become trapped in a paranoid feedback loop, one that simultaneously confirmed and amplified his suspicions and opened up his mind directly to Air Force Intelligence. That is not a situation that anyone can maintain for long without showing signs of serious strain, yet Bennewitz's ordeal lasted years and was to get even stranger before it was over. Even in the grip of paranoia, Moore would later recall, Bennewitz still talked 'an excellent story' and would do so to anyone who would listen. 'Most ufologists who listened to Paul based their own information on what he said, without investigating further.' The result was that Bennewitz's paranoid fantasies began to seep their way into the UFO underground. AFOSI were directly shaping what people thought about UFOs and were using Bill Moore to provide them with feedback. It was a textbook Psychological Operations scenario.

Keeping tabs on Paul Bennewitz and the flow of information passing through him wasn't all that Bill Moore did for his new 'employers'.[9] One piece of classic spy work made use of his fluency in Russian. American Intelligence assets within the USSR would routinely send postcards to UFO researchers in the US, apparently asking for information. 'Most of them were innocuous . . . but a few, like the ones I got, contained encoded bits of information on [Russian] facilities, weapons systems and defense.' On receiving one of these cards Moore would telephone an untraceable government number and read off its contents, at which point a voice would say 'thank you' and hang up. Each postcard contained only part of the intended message, which would be collated with others to make up the bigger picture. Once the contact had been made, Moore would send the postcard to a PO box in Washington DC without ever finding out what it was that he was passing on.

Sometimes Moore and, later, trusted colleagues whom he brought into the fold, would be handed documentation without knowing whether or not it was bogus. In early 1982 Falcon telephoned Moore to give him a 'recognition signal', a password, indicating that something would be coming his way. On 1 February a man approached Moore, said the password and handed him a manila envelope. It contained internal Kirtland security documents describing UFO sightings around Manzano in August and September of 1980. The documents were almost certainly fakes intended to back up Paul Bennewitz's stories,[10] but at the time Moore had no idea what kind of data he was being given access to.

In 1983 Moore was told that he would soon receive some important information. A bizarre wild goose chase followed, in which he was instructed to fly to airports all around the country until finally reaching a hotel in New York State. A courier came to his room with the inevitable manila envelope and gave him nineteen minutes to study its contents, during which time he was able to photograph the documents inside and read them into a tape recorder.

The papers purported to be a UFO briefing for President Jimmy Carter, who had once promised to reveal to the public everything the US government knew about the subject. Further references were made to Project Aquarius and the Majestic 12 (MJ-12) group mentioned in the fake AFOSI document from 1981. The message presented by these new documents was that a clandestine government agency, MJ-12, had been set up to engage with UFOs and their occupants, and to keep the reality of the ET presence a secret, since at least the early 1950s. These documents, which were later shown to other UFO researchers by both Moore and AFOSI, laid the groundwork for what would be the most devastating strike to the UFO community in its entire history. The alien seed planted by AFOSI was about to grow wildly out of control.

At the end of his 1989 MUFON presentation, Moore presented his own 'Estimate of the Situation' regarding the UFO subject. Some of his statements seem sensible to us now, others reflect the tangle of paranoia that was infecting the UFO community at the time, even its better grounded members like Moore. Moore told the audience that:

> – A highly advanced ET civilization is visiting planet Earth and is actively manipulating our awareness of its presence here.
>
> – Elements of at least two government agencies are aware of this and are conducting highly classified research projects. One of these projects has the data to prove that some UFOs represent somebody else's highly advanced technology.
>
> – USG counter-intel people have been conducting disinformation about the UFO phenomenon for at least forty years. High-level operatives in at least two agencies are behind it and are co-operating to some extent. They are creating false documents and gathering information on researchers and experiencers within the UFO community using informants.
>
> Why are they doing this? The disinformation provides security cover for a real UFO project that exists at a very high level and is known only to an elite few.
>
> – It is useful for diverting attention away from USG R&D projects.
>
> – It aids those groups, such as the Trilateral Commission, who are using the UFO phenomenon to try to bring about world unity in the face of an unknown threat from space.
>
> – It provides a convenient way to train counter-intelligence agents in deception and disinformation.

– It is a manipulation by the aliens themselves to gradually make human society aware of their presence.'[11]

If we disregard the comment about the Trilateral Commission, a cipher for the multiple New World Order conspiracy scenarios that were proliferating at the time, we are still left with the fact that, despite his own deep involvement in the web of deception spun by the intelligence services, Moore still demonstrated an unshakeable faith in the reality of the ET presence on Earth.

It might seem amazing that after almost a decade entangled with AFOSI Moore had not rejected the entire UFO scenario out of hand as an intelligence ruse. But he still believed and, more incredibly, so did his handler, Rick Doty. And now, seventeen years after he was forced out of the shadows, John and I were preparing to meet him.

NINE
RICK

'I am forever amazed at individuals such as these that think they can walk in and talk with one such as Rick and that he would bare his soul, "talking as a civilian", and tell them everything. It just does not happen that way'

Paul Bennewitz, letter to Christa Tilton.[1]

The Laughlin UFO Convention was now fully under way. John and I sat in the main foyer opposite a giant inflatable cartoon alien, eyeing passers by and wondering if Rick Doty was going to be among them. And if he was, would we recognize him? Greg Bishop warned us that at least two Rick Dotys had been identified in the past, though the one he had interviewed for his book was the one in the photograph. That didn't really help our search. As we surveyed the milling crowd, mostly retirees of a certain age, this being a weekday at a week-long conference, some of them would eye us back. Were they undercover agents, or were we making people paranoid?

Every so often, Bill Ryan would pass us by. He told us he was also feeling watched: as the convention's emissary from planet Serpo, Bill was drawing a lot of attention, and it was starting to bother him. 'There's an incredible amount of politics and paranoia going on here,' he told us on one flyby. 'Everybody is telling me who I can and can't trust and everybody has a theory about what's going on.'

Greg Bishop, who was booked at the convention to talk about Paul Bennewitz, had something to tell Bill, especially when it became clear that Rick Doty had been his key adviser on matters ufological. 'Be very careful Bill. If Rick is involved I wouldn't let yourself get drawn in too deep. You know what happened to Bennewitz. Serpo is either government disinformation, or it's a scam. One way or another you've got to make sure you don't end up fried like Paul.'

But Bill was resolute: 'Even if only 10 per cent of it is true and the rest is disinformation, then this is the biggest story in human history and I can't let that go.' With the interview requests already piling up, Bill was also enjoying his moment in the limelight. After all, how often do you get the opportunity to change the world?

So we watched and we waited. A presentation on submarine alien bases off the Puerto Rico coast drew a packed crowd, while Greg spoke to a less than full house. Seventeen years after Bill Moore's revelations, people still didn't want to know what was going on around them.

Paying close attention to Greg, however, was a curious group who had earlier caught my eye. They sat at the same table during each presentation, the one nearest the exit, and had done so now for a couple of days on the trot. Two hulking, well-built men in their thirties, both well groomed, one of them bearded with a long, neat ponytail under a large cowboy hat, sat on either side of a thin, weather-beaten man in his sixties. Grey-haired and steely-eyed, he wore a brown leather flight jacket and, like his companions, looked in impeccably good shape – in startling contrast to the majority of conference goers. I became convinced that they were military observers or intelligence agents; after all we'd learned it would be more surprising if there weren't any spies out there. Meanwhile we were looking for our own not-so-secret agent, Rick Doty.

Was he going to show up? Bill seemed to think so, as did Greg, but we had our doubts. Surely one of the most notorious, distrusted – even despised – characters in UFO history wouldn't just saunter into a hotel full of ufologists. He'd at least be incognito. With this reasoning as our guide, we began to give our prime Doty suspects ratings out of three. The man in the Hawaiian shirt and shorts? Hmm, a two. Perhaps Rick had grown a beard? A one. Was he tall or short? Maybe he was bald and was wearing a toupee in the photograph? Zero.

By mid-afternoon on our third day we were beginning to despair of Doty ever showing up. When we'd set out on our crazy mission we'd never expected even to make contact with the man, let alone get to meet him. It was looking as if he was to remain an enigma after all.

A pair of grey flannel trousers passed me at eye level, a conference badge clipped to the pocket: 'Rick Doty'. I jumped off my chair and called out. 'Rick!'

'Hi! Are you the film guys from England?'

Rick looked very much as he did in the photograph we'd seen. Mid-fifties, tidy, short, greying brown hair, wire-rimmed spectacles, a white and grey striped short-sleeved shirt with pens tucked into the breast pocket. He looked the picture of non-distinction, more civil servant than secret agent, more Bill Gates than Wild Bill Donovan – perfect.

'We thought you'd be here incognito – isn't it a bit risky you showing up at a UFO conference like this?'

'No,' he laughed in a high-pitched snicker, 'hardly anyone here knows who I am. And I haven't been to a UFO conference in a *long* time.'

Rick sat down and began to talk. He told us he was currently working for Homeland Security with a special interest in computing, and was also training as a lawyer. He was here, he repeatedly told us, as a private citizen with a keen interest in UFOs. Over the course of the next few days, 'private citizen' would become something of a mantra for Rick, spoken in a nasal, gently rising and falling singsong voice. 'When I first arrived at the hotel and was in line to pick up my room keys I spotted a DIA guy I knew a couple of people behind me. He looked surprised to see me and asked what I was doing here. I told him "I'm just a private citizen", enjoying the UFO conference like everybody else.'

'The Defense Intelligence Agency are at the conference?!'

'Sure, there are Intelligence people at every UFO conference. There's a few of them here, just keeping an eye on things, finding out what people are interested in, what they've seen, what they think they've seen. There's a contingent of Chinese UFO researchers here this year. The DIA guys are going to be *very* interested in what they're here for. But me, I'm just here as a private citizen.'

Rick looked up suddenly. Bill Ryan had appeared at our table.

'Rick! Great to see you, we'll talk later! Hi guys!'

'Hey Bill, what are you doing? Care to join us?'

'No thanks, I've been invited to a private screening of a video. It's an interview with a live extraterrestrial. I'd take you along but only six of us have been invited. See you later!'

Bill strode out of the foyer, clearly in his element, already settled into the the starry folds of the community's inner sanctum.

'I've seen that video,' said Rick, a touch of disdain in his voice. 'It's bullshit.'

He looked at us both seriously for a moment. 'You're not with Bill, are you?'

'With him? No, we're not travelling with him, if that's what you mean. We're here to film him, but we don't know him well.'

Rick looked relieved. He told us that he had some business to attend to, but would we like to meet for dinner later that evening? John and I agreed, doing our best to hide our glee.

Once Rick had disappeared from view we couldn't resist a high five, a pair of beaming disinformation fanboys. Holy shit! Rick Doty, ufological man of mystery, had just invited us out to dinner. What next? Recruitment?

Later that afternoon, we told Greg Bishop that we'd seen Rick and were going out for dinner with him.

'Oh you'll have a good time,' he told us, sounding a little nonplussed. 'He'll tell you some good stories, that's for sure. Make sure you pay attention, because each time you hear them they'll be just a little bit different.'

LIFE IN THE SHADOWS

The *Colorado Belle* is an old paddle steamer converted into a restaurant, or perhaps a restaurant converted into a paddle steamer, it's hard to say which. Here, over a classic American dinner of ribs and fries – and salad for John, a vegetarian – Rick regaled us with tales of espionage and derring-do. He told us that in the late 1960s he'd worked at Area 51, the ultra-secret Nevada desert airbase so often associated with UFOs; that in the 1980s he'd disguised himself as a Russian *babushka* on the streets of Moscow to tail a spy; he'd installed cameras in a painting to catch the Bulgarian Defence Minister up to no good with his manservant, and hidden microphones and transmitters in the aluminium bats of the Russian army's softball team. He boasted of his hacking skills, informed us how to evade a lie detector test ('clench your sphincter') and wowed us with descriptions of lasers that allow their user to see around corners – 'Oh I don't think that's classified any more' – and spy dust that leaves ultraviolet trails.

It was hard to tell whether these were things that had happened to Rick himself, or whether they were tales of the spy trade that had been handed down and passed around the community for all to share, especially when trying to impress new contacts. There's nothing in Rick's service records to suggest that he ever spent time in Moscow, or at Area 51, though if these were classified assignments then perhaps they wouldn't be on record. Actually it's quite hard to find out much at all about Rick Doty, and there's plenty more that he's just not telling.

Richard Charles Doty may have been born in 1950 in Barton, New York; some online sources state that he was born in Roswell,

New Mexico, which is a lovely thought, though almost certainly untrue. The Doty family did have strong ties with New Mexico, however, and UFOs are in the Doty blood.

In 1951 Rick's uncle, Major Edward Doty, a meteorologist by training, had been put in charge of the US Air Force's Project Twinkle, which since late 1949 had monitored sightings of strange lights seen near various sensitive military installations around New Mexico. The lights were mostly green fireballs that flew silently for great distances then fell vertically, burning up before they hit the ground. They appeared repeatedly over the Los Alamos complex and nearby Kirtland and Holloman air force bases, which both incorporated some of the most sensitive military research then taking place on the planet. Many of the witnesses to the fireballs were military, intelligence and scientific personnel who, despite their combined knowledge and experience, remained puzzled by the phenomenon, which most of them felt were neither flares, rockets nor meteorites. Meteorite expert Dr Lincoln La Paz, who reported two UFO sightings, one a green fireball, was convinced that the fireballs were artificial, classified American or possibly Russian technology, as was Leon Davidson, also based at Los Alamos at the time. But by the time a half-hearted attempt at setting up monitoring stations in New Mexico had got under way in 1950, the green fireballs seemed gradually to fade from view and, with Major Doty as its part-time caretaker, in 1951 Project Twinkle was finally closed down.[2]

Uncle Ed's role in UFO history aside, as a child Rick had little or no interest in flying saucers, though his brother was fascinated by them (a source of much amusement to young Rick). In September 1968 he joined the US Air Force, receiving his basic training at Lackland AFB in Texas before being moved to Sheppard AFB in the same state. And that's about all that we can say for certain, as from here his story quickly gets hazy.

Doty's military records show that he remained at Sheppard as a security policeman until being sent out to Phan Rang airbase in

Vietnam. But did something remarkable happen to him in the meantime? Greg Bishop describes the incident in *Project Beta*, and Doty presents it again in a self-published book called *Exempt from Disclosure: The Black World of UFOs*, written by Robert Collins, a former Air Force physicist and, like Rick, one of Bill Moore's 'aviary' of military insiders with an interest in UFOs.[3]

In the book Rick describes how, in July 1969, he was assigned a top-secret duty at a secret facility near Indian Springs AFB in Nevada. Here he stood guard outside a huge hangar, 3,500 feet by 4,000 feet and 100 feet high, which he was told housed 'experimental aircraft'. One day the hangar doors opened and a large flying saucer was pulled out by a tug. It sat on the runway for an hour, being fussed over by men in lab coats, but never took off – a routine performed several times during Rick's forty-five days at the site. One afternoon a civilian Doty refers to as Mr Blake asked him what he knew about flying saucers: 'What if I told you that the object you saw today was a real flying saucer from another planet, then what would you say?' Mr Blake told the puzzled young guard that one day he would learn the truth about the craft he had seen.

The story – we might call it Rick's origin myth – is told somewhat differently in *Project Beta*. This time the incident takes place explicitly at Area 51. A large, dull-black saucer-shaped craft is pulled out on to the runway, where it silently starts up and, surrounded by a blue electrical corona, lifts about 200 feet from the ground. Doty claims to have witnessed several such tests, once overhearing a technician say, 'I think we can get this thing out of the atmosphere.' On another occasion the commanding officer – not Mr Blake – tells Doty that the craft was 'what is generally known as a UFO, and it's not one of ours. It's on loan.' As in the other version of the story, Doty is told that one day he will learn a lot more about the craft.

Doty's account is typical of the sort of military tall tale that often crops up within UFO literature. But at this point we reach something of a brick wall. Are these tales fictions? Does the US government really possess alien spacecraft? Or could they be staged events, used to test a subject's ability to stay cool when confronted with the unexpected, or to keep a secret? Then, if the subject does blab about flying saucers, no real harm has been done and they will be regarded as just another UFO kook among many. Rick, however, didn't strike us as just another UFO kook.

After Vietnam, Doty was stationed for two years at McChord AFB in Washington state, then served three years as a gate guard at Wiesbaden, West Germany. In 1976, aged twenty-six, he was moved to Ellsworth AFB, South Dakota, and it's here that he probably got his first glimpse – Area 51 flight tests notwithstanding – of the true power of the UFO. Ellsworth at this time housed both strategic bombers and Minuteman Intercontinental Ballistic Missiles (ICBM), making it a critical component of America's Cold War arsenal. In November 1975 missile silos in neighbouring states had been plagued by UFO sightings and a spate of bizarre cattle mutilations. These were all recorded in US Air Force documentation, making them likely subjects for base gossip and speculation.

Around this same time, the *National Enquirer* became perhaps the most outspoken and reliable source of information about UFOs, unlikely as it may seem. In late 1974, five years after the Air Force had closed Project Blue Book, its official UFO investigation, the *Enquirer* set up a 'Blue Ribbon' panel on UFOs in conjunction with civilian UFO research organizations APRO and[4] the National Investigations Committee on Aerial Phenomena (NICAP) and scientists including J. Allen Hynek. The UFO groups passed their most interesting cases to the *Enquirer*, who offered to fund further research into the incidents if the panel deemed it necessary. At the same time, the *Enquirer* was offering a cool $1 million to anybody who could prove that UFOs did indeed come from Outer Space, and a not inconsiderable $5,000–10,000 to witnesses in what it declared to be the year's best cases.

In February 1978, just three months after Spielberg's *Close Encounters of the Third Kind* became a runaway sensation, an anonymous letter postmarked Rapid City, South Dakota arrived at the *National Enquirer*'s Florida office. The letter appeared to be from the commander of the 44th Missile Security Squadron and described a security breach the previous November at one of the Minuteman silos near Ellsworth. An accompanying Air Force report form also detailed the incident.

According to the letter and document, a two-man Security Alert Team had been despatched to investigate the breach. Arriving at the silo, they were confronted by a humanoid in a green, glowing metallic suit and helmet. The being fired a weapon at the men, dissolving one of the men's rifles (as in *The Day the Earth Stood Still*) and seriously burning his hands. The other guard spotted two more figures: he shot one of them through the arm and the other through the helmet. Seemingly unhurt, they disappeared over a hill and entered a thirty-foot flying saucer that took off at great speed. It was later discovered, the letter reported, that the missile in the silo was missing its nuclear components.

Three *Enquirer* reporters, including Bob Pratt, who later became a respected UFO investigator in his own right, headed to Ellsworth to do some digging around. As they did, the story began to crumble around them. Having spoken to everyone alleged to have been involved in the incident, they discovered that the whole thing was a hoax – the Air Force report form was a well-constructed fabrication. While the named men all served at the base, their roles were not as described in the letter, some had different first names to those used in the report and some of the locations had been mixed up. In all, the team found twenty errors and the *Enquirer* never ran the story. But the Ellsworth document did leak into the UFO community to be touted as evidence of ET interest in Amercia's defensive capabilities, later piquing the curiosity of Paul Bennewitz. Bob Pratt didn't publish his own account of the hoax until 1984.

Who was behind the Ellsworth letter? Although Rick Doty was serving at Ellsworth at the time, he denies any involvement in the affair, stating that he wasn't recruited into AFOSI until spring 1978;

but it does sound like the sort of thing AFOSI would later get up to with Bill Moore and Paul Bennewitz. Despite being exposed publicly as a hoax, the Ellsworth papers are a textbook study in disinformation. They tell an absurd story in such a way as to make it seem, at least initially, plausible to the group being targeted, in this case the *National Enquirer* and the UFO community. At first glance the documents looked real and the story checked out, leading the *Enquirer* to waste several-thousand dollars and a couple of hundred man hours investigating the incident.

So why would AFOSI want to disinform the *Enquirer*? To diminish the publisher's enthusiasm for UFOs? Bob Pratt later said that its publisher, Generoso Pope Jr, spent tens of thousands of dollars sending him all around the world chasing UFO stories, even though they sold fewer issues than celebrity stories. Pratt thought Pope was a genuine believer in UFOs, though others have suspected an intelligence connection. Pope spent a year being trained in psychological operations by the CIA in 1951, the year before he bought the *Enquirer*. He was also a close friend of Nixon's Defense Secretary, Melvin Laird.[5]

The Ellsworth hoax may have served as a temporary distraction from rumours about the nearby ICBM silo intrusions of 1975, or acted as a foil to the peace-loving aliens of *Close Encounters*, whose conclusion took place at Devil's Tower, coincidentally (or not), just a few miles away from Ellsworth. The film had sparked a huge resurgence of public interest in UFOs, leading to a renewed clamour for a fresh Air Force investigation, something the US Air Force were keen to avoid. Or perhaps it was all nothing more than a disinformation training exercise for AFOSI's new recruits.

Even if Rick Doty wasn't directly involved, we can be sure that he would have heard about it. By March 1978 he had been enrolled for six weeks' training at the Non Commissioned Officers Academy, and by May 1979 he was stationed at Kirtland AFB with AFOSI. The operation against Paul Bennewitz would begin within a few months of his arrival.

THE AQUARIUS PROJECT

Spending time with Rick it was easy to see how he had gained the trust of Paul Bennewitz and Bill Moore. Sure, some of his stories didn't ring true – there was something about the way they tumbled from his mouth that felt a little second-hand – but it didn't matter, this was Rick Doty we were talking to, and hey, what's a little disinformation among new friends?

Sated with hormonally ravaged beef – or lettuce in John's case – the three of us decided to go and find Greg Bishop. We asked Rick what he'd thought of Greg's book *Project Beta*, seeing as he played such a key role in it. It was a good book, Rick felt, but like all good disinformation it contained some accurate information and some that was inaccurate. A little unexpectedly, he also expressed his disappointment that it hadn't sold as well as he'd hoped it would. Rick was clearly a man with keen commercial instincts.

Generally, however, books were among the things Rick wasn't too eager to talk about. He largely disowned *Exempt from Disclosure*, stating that he hadn't wanted his name on the cover, although it's still there on the second, revised edition. And, although he was reluctant to discuss it with us, back in the early 1980s Rick had been involved in another putative book project, one that never saw the light of day but might provide a glimpse into Rick's personality during the time that he was handling Moore and Bennewitz.

By late 1981, Doty and Moore's relationship was becoming more firmly established. The flow of information was a two-way process: Doty would offer titbits to Moore – project codenames, hints at some of the personnel involved in the UFO cover-up – while Moore would provide Doty with the latest information from inside the UFO community. Moore was also in close contact with Bob Pratt, who by now had gained the community's trust and respect through his articles for the *National Enquirer*. It's unclear exactly who instigated the project – Moore and Doty aren't talking and Pratt is now dead – but it's likely that Doty and Moore first discussed the idea of publishing a book containing all the information that they were sharing, a follow-up to Moore's successful *The Roswell Incident*. The idea was that Doty, credited as 'Ronald L. Davis' would be the insider source, while Moore and Pratt would write it. The material intended for the book was essentially to be the same as what they were feeding to Paul Bennewitz in the form of faked government documents, and would detail the ongoing secret UFO investigation set up following the Roswell crash.

Bob Pratt recorded all his telephone conversations with Moore and his tapes reveal that Moore was initially insistent that the book should be presented as non-fiction, but Pratt was uncomfortable with the idea, given the lack of evidence for the material that Moore, via Doty, was presenting him with. Reluctantly, Moore agreed to a fictional account, first called *Majik 12*, then renamed *The Aquarius Project* after the faked documents given to Moore by Doty. The book, as plotted by Pratt and Moore, would tell the story of D, a patriotic American soldier who returns from a devastating tour of duty in Vietnam feeling betrayed by his country. D is recruited into the intelligence field, where he is tasked with feeding false UFO information to one Dr Berkowitz (a thinly disguised Paul Bennewitz), before being called in to investigate a case of an unknown object tampering with a nuclear missile in a South Dakota silo (a play on the Ellsworth hoax). D gradually discovers that there is a high-level, super-secret UFO programme buried deep within the US government. This is Project Aquarius, and the organization that acts as its custodian is Majestic, or MJ-12.

As D is drawn deeper into the UFO conspiracy he learns that historical figures including Jesus Christ, Muhammad and Adolf Hitler were all in the control of extraterrestrials. Reflecting information supplied to Moore by Doty, D learns that there are three alien species interacting with the Earth: the beatific Nordic-looking humanoids (like George Adamski's Orthon) had originally seeded human life here and were discreetly guiding our development; the malevolent Greys were abducting humans and mutilating cattle as part of a genetic harvesting programme, while a third species wanted to plunder Earth's natural resources. The US government knew all about these aliens and, through MJ-12, kept tabs on them, occasionally negotiating with them in exchange for advanced technologies. But MJ-12 was ultimately powerless to stop the aliens, hence the need for a cover-up. How does a government tells its citizens that their genes and the planet's resources are in alien hands without inducing a state of total chaos? After getting too close to the truth, our hero D decides that the people deserve to know what's really going on. He starts leaking genuine UFO material to researchers – like Bill Moore and Bob Pratt – but is assassinated by MJ-12. His body is handed over to one of the alien races and flown back to their planet in a downbeat twist on *Close Encounters of the Third Kind*'s grand finale.

Was this all a fiction or did Doty really view himself as a martyr to the truth about UFOs? More likely, this is the impression that Doty wanted Moore to have of him. He and Falcon had originally presented themselves as insiders who disagreed with the government's UFO policy, and had promised to provide Moore with real UFO secrets. In some ways D, the book's martyred hero, is a composite of both Doty and Moore, a man willing to risk his integrity, his soul, perhaps even his life, to get to the truth. It's a role that Moore projected himself into for his revelatory 1989 MUFON lecture; he does genuinely seem to have felt that he was doing the right thing in colluding with AFOSI, and also to have believed that some of the information they were feeding him was true.

But what about Doty? In a letter written in 1989, shortly after Doty's role in UFO-related disinformation was made public, he wrote:

> Whether Earth has been visited in the past by visitors from other planets I simply don't have enough information to make a personal decision. If I should base my decision on information that I had

> access to during my Government Service, I would have to say, yes, Earth has been visited. However, I am not 100 per cent certain that the information I had access to was entirely accurate.[6]

Compare this with his 2006 *UFO Magazine* article: 'In early 1979 . . . I was briefed into a special compartmented program. This program dealt with United States government involvement with extraterrestrial biological entities. During my initial briefing I was given the complete background of our government's involvement with EBEs.' Which Richard Doty should we believe and, more critically, what does *he* believe?

THE YELLOW BOOK

John, Rick and I met Greg Bishop and his fiancée, Sigrid, at the Muddy Rudder, a typical American dive bar – low lighting, harsh neon advertising and TV sports – that extended the riverboat theme so popular in Laughlin. Greg and Rick seemed to enjoy a friendly, if spiky relationship. Greg made it clear that he didn't believe much that Rick – or anybody else – told him, while Rick humoured him with the occasional barbed response. Beers were drunk. Quite a few beers, in fact. Even Rick was drinking, the only time we ever saw him do so.

Initially, UFOs weren't a topic of our conversation, but as lips loosened so did our inhibitions and eventually we had to talk about why we were all there. While the rest of us maintained our agnosticism on the matter of ET visitation, Rick stuck to his guns. They had been here, he asserted, the US government knew about them, and they had the technology to prove it.

Loudly, maybe too loudly, I expressed my incredulity. At first Rick took my suspicions on the chin, like a man who has had to defend his corner before. Becoming more garrulous, I pressed harder. Unexpectedly, Rick became defensive: 'I know the technology is real because I've handled it,' he told us.

'Handled what?'

'Alien technology. A crystal . . . some kind of holographic device. They call it the Yellow Book. I've held it. It's a rectangular slab of crystal, like a hardback book. There are indentations on the front, one on each side, and you put your thumb or finger there. Then you see things.'

'See what?'

'Well, I only saw words. Other people saw images . . . I'm telling you it's real.'

'How do you know it wasn't just some kind of futuristic military Palm Pilot technology?'

'This was alien technology, simple as that. And if you had been there, you would know that, too. I don't care whether or not you believe me. I was there.'

And that was the end of the conversation.

Now at this point I'd like to say that I'm a good judge of character and that I know Rick was telling the truth. But I can't say that. I was drunk, I think Rick was drunk, and I know that everybody else was drunk. What I can say is that the defensive stance taken by Rick during that exchange, the high-pitched whine creeping into his voice, the resigned shrug of the shoulders, all felt unguarded and genuine. At that moment, I believed him.[7]

From that point on I felt that Rick really did believe that we had been visited. Why he believed that – how he believed that – was another matter, and something we can't know for ourselves.

Perhaps Rick has told the same lie for so long that he has come to believe it himself. But I can't help feeling that somewhere in the maze of stories that gush from him like steam from a geyser – the briefings, the flying saucer test flight, the Yellow Book – something had happened to him that had left him convinced. Rick *was* shown something, or was at least *told* something by people he trusted – friends and superiors in the military. And, perhaps in the same way that his spy stories felt like second-hand accounts of other people's war stories, so some of the UFO stories may not be his own. But that doesn't necessarily make them less real. There's a strong culture of belief in UFOs and ETs within the US military, and there are people serving there who joined the military hoping to get a glimpse into the ufological Aladdin's Cave. Maybe there is something truly alien hidden away there, or maybe all these artefacts are just props in the disinformation game. And maybe Rick had seen them.

However, whether we believe Rick Doty or not, we should also never forget that he was trained to deceive; he did so during his time with AFOSI at Kirtland and, twenty-five years later, he was still involved in some extremely rum goings-on. And they were about to get significantly rummer.

TEN

BEEF, BUGS AND THE ALIEN UNDERGROUND

Article 3: The Parties undertake to notify each other immediately in the event of detection by missile warning systems of unidentified objects, or in the event of signs of interference with these systems or with related communications facilities, if such occurrences could create a risk of outbreak of nuclear war between the two countries.

USA/USSR Agreement on Measures to Reduce the Outbreak of Nuclear War *(30 September 1971)*

The next morning, nursing slight hangovers, John and I met in the Flamingo's Aviary Lounge as the conference stirred into life, fuelled by the coffee that was permanently on tap. The ufologists needed it; talks began at 9 a.m. sharp and continued until 5 p.m. with only a break for lunch. On today's agenda was an Englishman telling us what was going to happen in December 2012; another Englishman discussing the UK military conspiracy to cover up his crop circle research, and a presentation about a Utah ranch that hosted a malevolent Native American spirit being, an interdimensional wormhole and, of course, UFOs.

Rick Doty appeared at about 10.30 a.m. He looked happy to see us and sounded chipper, despite having been woken up at 5.45 a.m. by one of his DIA contacts at the conference. The DIA man wanted to talk to Rick, so they'd met in the hotel car park, got in a hire car and driven to the outskirts of town before beginning their conversation, which we weren't authorized to hear about. Rick had another such meeting that afternoon, though he insisted that these were informal exchanges and reminded us that he was at the conference as a 'private citizen'.

We chatted about the day's presentations. What about that ranch in Utah, known as the Skinwalker ranch? During the early 1990s the owners had found several of their cattle mutilated – drained of blood, their sexual organs neatly sliced off, ears, sometimes tongues removed, their rectums cored out, all with no signs of struggle. The bizarre pattern echoed similar mutilations that had plagued ranchers during the 1970s, right down to the appearance of Bigfoot-like creatures near the dead animals, and odd lights in the sky.

'Skinwalker? That was a hoax, wasn't it?' Rick was dismissive. 'Really guys, 90 per cent of all this stuff is bullshit. The truth is that there are ETs, they came here for a while, we kept two of them in captivity, got some of their technology and then they left in 1965, taking the Serpo team with them. That's what happened. All the rest is just baloney, especially all that abduction crap.'

'Really? All of it?'

'Well there was one abduction that puzzled us: a woman who was taken from around Four Corners – that's the area where New Mexico, Colorado, Utah and Arizona meet. There's nothing but rocks and snakes out there. She was taken one night and even left her baby behind in the vehicle. Her husband was in the Air Force and asked AFOSI to look into it.' Rick laughed, then paused, becoming pensive for a moment, as if suddenly confronted by the great void of unknowing. 'Yeah, we never did work out what that was. You know, I have a great telescope at home, it's connected to my computer. I like to look at Deep Space objects with it and wonder what's out there. And something *is* out there, I know that much. If I try to work it out, I reckon I know about 35 per cent of the big picture. The rest . . . well, somebody knows what's going on – that I'm sure about.'

As we talked, a familiar figure approached the table. It was the weather-beaten man, the one I'd noticed with the two burly escorts during the lectures. He walked past our table, made eye contact with Rick and gave him a deliberate, knowing nod. We all pretended it didn't happen. Inside, however, I was freaking out. Was he one of Rick's DIA contacts?

Discussing the cattle mutilations at the Skinwalker ranch brought us to one of the most bizarre aspects of the Paul Bennewitz saga, and one with which Rick was directly involved. It also laid the foundations for one of the most potent and paranoid elements of the new UFO mythology: an alien base hidden inside the Mesa overlooking the tiny town of Dulce, New Mexico, a small Jicarilla Apache reservation town about 200 miles north of Albuquerque, near the Colorado state line.

A week after the conference was over, John, Greg Bishop and I paid Dulce a visit. The first thing that strikes you about the town is how small it is and how poor it is. The second thing that strikes you is the beauty of the surrounding landscape. Lush green valleys roll out between spectacular, snow-capped rocky peaks and, dominating the landscape, the huge glacier-like wall of the Archuleta Mesa. I had expected the Mesa to be something that could be easily encapsulated, and so easily dismissed as the home of an alien underground base. But really it's a landlocked island, about twenty-five miles long, ten miles wide and three hundred feet high in places. You could fit several alien bases in there.

The other thing that strikes you about Dulce is the highly incongruous Best Western Jicarilla Inn and Casino, a modern, fairly luxurious hotel sitting off the highway at what is probably the heart of town. Two huge bronze horses rear up in front of the entrance; not the friendliest of welcomes, but the staff inside were pleased to see us, and the gift shop even sold UFO t-shirts: 'My friend got probed and all I got was this lousy t-shirt.'[1] So how on earth had such an inconspicuous place gained such a sinister reputation? The answers lie in the mind of Paul Bennewitz, and in a spectacular AFOSI illusion.

THE ORGAN SNATCHERS

On 20 April 1979, exactly a month before Doty began AFOSI duty at Kirtland, New Mexico, Senator and former moonwalking astronaut Harrison Schmitt convened a very unusual conference in Albuquerque. Present at the all-day meeting were ranchers and law enforcement officials from New Mexico, Colorado, Montana, Arkansas and Nebraska, all seeking answers to one question: who, or what, was killing and mutilating their cattle? Also in the audience were Paul Bennewitz, and, most likely keeping a low profile, representatives from Kirtland AFOSI.

Many ranchers and local police were convinced that humans were responsible for the deaths, perhaps a cult of witches or satanists, although an autopsy on a cow at Kansas State University showed that it had been attacked by animal, not human, predators. The ranchers didn't buy this explanation, however, and by late 1974 they had organized into armed patrols, ready to catch the mutilators bloody-handed. They found nothing. Meanwhile the deaths continued, fanning out across America's ranching states and eventually making the national news.

Official investigations in Oklahoma and Colorado both ruled out human involvement, once again blaming natural death and predation, but that didn't stop the apparent mutilations – or the rumours. As the wave of fear spread, so the stories grew weirder, with reports of strange lights and helicopters seen near mutilation sites. In 1974 a Colorado newspaper reported that a sheriff had found an army bag containing surgical gloves, a scalpel and a bull's penis at a mutilation scene.[2] The most puzzling aspect to the mutilations was how the perpetrators could enter and exit a site silently, without leaving any traces: it was, therefore, only a matter of time before someone made the UFO connection.

In 1975, Montana was hit by a tsunami of strangeness. There were cattle mutilations, sightings of egg-shaped aircraft, unmarked helicopters, UFO sightings over its Air Force ICBM silos and, most puzzling of all, appearances by bigfoot-like creatures that appeared impervious to gunfire. Suspicions mounted that this was about more than cultists. All the investigative legwork was being done by local sheriffs; the regional military and Air Force didn't seem to know, or care, what was going on, and the federal government refused to get involved. Things were threatening to get out of hand as ranchers formed armed posses to protect their herds. Fear spread through the community, ready to explode into panic at any moment; sheriffs were called in to calm students at a Montana high school after it was rumoured that they were to be the mutilators' first human targets. It was exactly the kind of irrational hysteria that the CIA's Robertson Panel had warned of back in 1953.

By the late 1970s New Mexico was in the grip of its own mutilation wave; ranchers and local sheriffs had begun taking pot shots at aircraft flying over their property and a major incident was only a gunshot away. The Albuquerque conference of 1979 was Senator Schmitt's attempt to gain control of the situation before humans, as well as cattle, got hurt. It was here that Paul Bennewitz first encountered Gabe Valdez, a highway patrolman based in Dulce. The town had been plagued by cattle mutilations and UFO sightings since 1975, and Valdez had taken it upon himself to investigate on behalf of local ranchers.

John and I met Gabe Valdez at his comfortable home outside Albuquerque the week after our own conference. Now retired, the serious but genial Mexican-American was still fascinated by the mutilation mystery; in 1997 he'd written a detailed report on the subject for the National Institute of Discovery Science, a paranormal research organization headed by Las Vegas hotelier Kevin Bigelow, who had also funded the investigation into the Utah Skinwalker ranch we'd heard about at Laughlin.

Back in the 1970s, Valdez's investigation had focused on the ranch of Manuel Gomez, whose cattle had suffered particularly badly; alongside dead and mutilated animals he'd found caterpillar tracks, bits of paper, measuring tools, syringes, needles and a gas mask. One site was covered with radar-reflecting chaff, some of it stuffed into the dead cow's mouth. Some of the animals had broken bones and what appeared to be rope marks on their limbs, suggesting that they had been hoisted up then dropped back on to the ground. Whoever was doing this to the cattle, they were organized, and human.

As in Montana, the mutilations coincided with a rash of UFO sightings, and Valdez and his fellow officers had several close shaves with strange flying machines. On one occasion Valdez's team cornered an orange light in a field; as they approached, the light went out. Then, although they could see nothing, they heard a muffled sound like a lawnmower engine pass over their heads. Another time Valdez and two colleagues ducked beneath an object that he described as disc-shaped, rotor-less, and dazzlingly bright. Valdez described the noise it made as it flew over them as 'put-put-putting' or 'ticking' – hardly the sound of advanced alien technology.

The intensity of the mutilations around Dulce had also drawn the interest of Howard Burgess, a retired Sandia Labs scientist, and one night in July 1975 he, Valdez and Gomez, the ranch-owner, followed a hunch to see whether or not humans really were behind the attacks. By shining an ultraviolet lamp on to the backs of a sample of a hundred cattle, the trio found that some of them had been marked with a substance that showed up only under ultraviolet light. The marked cattle were all between one and three years old and came from a particular breed: the same one as the dead animals found on Gomez's land. Gomez quickly sold off all the animals that fitted the mutilation profile.

Based on their discovery, the three men pieced together a *modus operandi*. The selected cattle were being marked with water-soluble

UV paint containing potassium and magnesium; this suggested that the marking was done quite soon before the operation was to take place. Under cover of darkness, whatever aircraft that the team employed would fly over the region and identify the marked animals using a blacklight beam that was invisible to any observers. The selected animal was tranquillized, probably from the air with a rifle, and then the mutilation operation was performed, either on the ground or, as Valdez believed, after the victim had been hauled to another location – hence the rope marks on some of them. Valdez told us that the mutilators had their surgery-cum-laboratory in one of many disused mine shafts on top of the Archuleta Mesa. Their grisly work over, they then returned the animal to the pastures, mutilated and drained of blood, to be found by an unlucky rancher.

Although Valdez's enthusiasm for the subject was clear, so was his anxiety; at one point he hinted to us that he had found the entrance to a military facility on top of the Mesa, but he quickly backed away from the subject when asked for more information. Valdez clammed up like this a few times, as if he felt he'd said something he shouldn't. We later learned that there was good reason for his hesitance: while conducting his investigations back in the 1970s he'd become convinced that he was under surveillance, a suspicion borne out by the discovery of a microphone 'bug' in his telephone handset.

What Valdez did tell us was startling enough: he thought that the military had been flying helicopters into the Dulce area from Fort Carson, an army installation about 300 miles to the north, near Colorado Springs, and using the Archuleta Mesa as a staging post for their mutilation missions. He also suggested that 'real' UFOs – flying saucers – were being flown alongside the mutilators' own aircraft, perhaps by another government agency, in order to obfuscate matters further, or at least to confuse the locals. That one confused us, too.

So what were the mysterious 'ticking' aircraft that Valdez and others saw and heard flying over their heads around Dulce? Could they be the fabled black, silent helicopters of conspiracy lore? As a mechanical myth of the technological age, the silent chopper is almost as potent as the UFO, and has always been considered just another phantom of the paranoid classes. If the silent helicopters are real, the argument went, then why are they never used in warfare, or over urban areas, where they would be most useful? The answer is: they were, we just never knew about it, until recently. The silent helicopter has now been revealed as not only a reality, but one that was flying as long ago as 1972. This was the Hughes 500P, the P standing for Penetrator, an aircraft known by the few who flew it as 'The Quiet One'.

The Pentagon's Advanced Research Projects Agency (ARPA – now DARPA, the Defense Advanced Research Projects Agency) had been trying to develop a silent chopper since 1968, using the Hughes 500, a light observation helicopter, as its basis. It was the CIA who took the results into the field, buying two for its Special Operations Division Air Branch and giving them to Air America, their infamous 'cut-out' company that flew covert missions over south-east Asia.

Ironically, given the usual arguments against the helicopters' existence, the Quiet One arose from the Los Angeles police force's need for an urban helicopter that would cause less of a commotion as it went about its business. In its initial modification, Hughes doubled the number of blades on the 500's tail rotor from two to four and arranged them in a scissor-like formation, cutting its noise emissions by half. When ARPA heard about the LAPD's copter, they realized that a quiet chopper would be extremely useful to them and stumped up the money for further research.

The 'whup-whup-whup' sound of a helicopter is caused by 'blade vortex interaction' – the blade tips whacking the miniature tornados caused by their high-speed rotation; Hughes found that by adding an extra blade to the main rotor and altering the blade tips they could almost eliminate this effect. The 500's exhaust was also fitted with a muffler, the air intake was given a baffle and the entire body covered with lead and vinyl pads. The result, while not totally silent, changed the aircraft's sound signature enough so that it was no longer recognizably that of a helicopter. And the Quiet One wasn't just near-silent, it was also nearly invisible: an infrared camera allowed the craft to fly and land without lights, though this feature, the height of military technology at the time, suffered numerous teething problems.

The Quiet One was test flown at Area 51, and in California, possibly causing a few UFO reports of its own. Then, in 1972, it was ready for action: the CIA's two Hughes 500Ps were flown to a secret airstrip deep in the Laos jungle. Nobody was to know that the Quiet Ones existed; photography was forbidden and the choppers had their own special hangar to hide them from nosey satellites and reconnaissance aircraft. They were incredibly quiet; soldiers stationed at the base said that when a Penetrator flew overhead it sounded like an aeroplane passing in the distance, an unnerving sensation when you could see it right in front of you. Despite their almost magical abilities, however, the Quiet Ones didn't fare terribly well in the field. One flew a single successful mission to plant a wiretap behind enemy lines in December 1972, but the other was destroyed during a training exercise. The surviving chopper was returned to Edwards AFB in California where it was allegedly dismantled.

The Quiet One's paper trail ends with the Pacific Corporation of Washington DC, a CIA front, and no records exist for further Hughes 500Ps, or any other 'silent' helicopters. This doesn't mean that they weren't built – they could have been hidden within the government's ever-expanding black budget – but nor is there any proof that they were. What the story of the Quiet One does tell us is that the technology for such a craft was fully functional by late 1972, and it doesn't require a great leap of faith to imagine that their descendants might have been flying over Montana and New Mexico by 1975. While we can't prove a causal link between the mysterious helicopters seen in the region and the mutilations, Gabe

Valdez and his men were buzzed by *something* that night in Dulce, and *someone* was leaving gas masks and other items of military paraphernalia near mutilated cattle. The question remains: who – and *why*?

MAD COWS AND PEACEFUL BOMBS

As concluded by the 1980 FBI report following Harrison Schmitt's conference, it's likely that some of the alleged cattle mutilations were natural deaths and predator attacks, transformed into something more sinister by the ranchers' anxieties. However, if we don't accept that predators were marking cattle with UV paint, flying helicopters over ranching land and leaving debris in their wake, and if we rule out extraterrestrials, demons and other non-human entities, then we are left with people.

In the early days of the panic, ranchers and the press were keen to lay the blame on cultists: satanists, wiccans or some kind of perverts. Unconfirmed witness reports described robe-wearing humans passing through fields or alongside roads, but no one ever seems to have stopped to speak to these sinister weirdos, or watched where they were going, and this avenue of investigation soon proved a dead end. The inaccessibility of many of the mutilation sites, coupled with the odd lights and helicopters seen in connection with the animal deaths, soon led investigators to seek answers elsewhere.

During the 1975 mutilation wave in Montana, locals connected the cattle deaths to the mystery lights seen over ICBM silos in the same period. Might they be, at least in some instances, the same aircraft that were put-put-putting around over Dulce? Were quiet helicopters with unusual lighting configurations being flown over nuclear missile silos in Montana, perhaps to test the security arrangements at the sites and assess military personnel's response to mysterious aircraft? Soldiers at the sites were allegedly instructed not to fire at the craft or even shine spotlights on them – odd orders if they really were of unknown origin. Alternatively, the intrusions may have been designed to test the target-acquisition capabilities of the Safeguard anti-ballistic missile system recently constructed in Nekoma, North Dakota. This complex, the only part of a proposed nationwide defensive network to be completed, was dismantled in 1976 – perhaps it hadn't done a very good job of spotting these friendly phantom craft? Either possibility, or another like it, would explain why local law enforcement had been kept out of the loop by the military – they simply had no need to know.

Although they were well reported in the local papers, it would be four years before the Montana intrusions were mentioned, briefly, in the national press, and they did so only following a leak to the *National Enquirer* in 1977.[3] Soon afterwards it looks as if an attempt was made, via the faked Ellsworth documents sent to the *Enquirer* in 1978, to identify the events with ETs and UFOs, perhaps a ploy to keep the nationals from investigating further. For reporters on the major papers, taking an interest in the ICBM incidents would have allied them with the kooks, the UFO conspiracists and, worse, the *National Enquirer*.

We may have some idea how mutilations were done, but we still have to ask why? ET genetic experiments aside, the most plausible explanation for the phenomenon has its roots in epidemiology. A number of researchers have raised the possibility that the mutilations were part of a clandestine study or experiment. The parts of the animal that are usually removed by the mutilators – lips, tongue, anus, udders and genitalia – are those most prone to contamination and infection. These are the soft, squishy parts where food goes in and comes out of the animal, where they are most likely to ingest and excrete bacterial, viral or chemical agents and where humans are most likely to look for them. So what might *they*, whoever *they* are, be looking for?

Radiation is one possibility, particularly around Dulce, which has a unique place in America's atomic history. A clearing in the Carson National Forest, about twenty-five miles south-west of Dulce, is marked with a small plaque reading:

SITE OF THE FIRST UNITED STATES UNDERGROUND NUCLEAR EXPERIMENT FOR THE STIMULATION OF LOW-PRODUCTIVITY GAS RESERVOIRS. A 29 KILOTON NUCLEAR EXPLOSIVE WAS DETONATED AT A DEPTH OF 4227 FEET BELOW THIS SURFACE LOCATION ON DECEMBER 10, 1967.

The detonation, part of the 'Plowshare Program' for the peaceful use of nuclear explosions, did succeed in part, creating an eighty-foot wide, 335-foot high, natural gas-filled cavity. Unfortunately the gas was rendered dangerously radioactive by the explosion and had no commercial value, so the site was sealed up for eternity. Perhaps the mutilating agency was looking for signs of environmental damage caused by radiation from the Gasbuggy test leaking into the surrounding area?

More recently, molecular biologist Colm Kelleher has connected the cattle mutilations to the spread of prion-related diseases, including Bovine Spongiform Encephalitis (BSE), aka mad cow disease. Kelleher suggests that the mutilations can be tied to outbreaks of Chronic Wasting Disease (CWD) in wild deer and elk, and that CWD- and BSE-causing prions have successfully leapt the species barrier from wild deer to farmed cattle, and so into the human food supply. Kelleher proposes that the origins of this prion epidemic can be traced back to the US government labs at Fort Detrick and Bethesda, Maryland, where dozens of human brains infected with the fatal neurological illness *kuru* were stored from the late 1950s onwards.[4]

Both of these epidemiological explanations answer one of the questions most frequently raised about the mutilation phenomenon. If the government, or some other agency, is behind the deaths, why don't they simply buy and breed their own cattle on which to experiment? The answer is that they need to sample the same animals that will end up in our burgers. But why do they leave the bodies behind? This is a tougher one to answer. Perhaps it's too difficult to dispose of the bodies; perhaps it was hoped that the ranchers might be able to claim insurance on the expensive animals; perhaps the mutilators thought that the ranchers would be too frightened to investigate further; perhaps once the initial rumours of satanic cults and bloodthirsty aliens began to circulate, the mutilators felt that the confusion sown by their strange acts was working to their advantage.

Really, we just don't know, but the mutilations continue to take place in both North and South America to this day. Wherever there are beef herds, it seems the mutilators are never far away, and so are the UFO reports.

HOW TO BUILD AN ALIEN BASE

By the time that Gabe Valdez met Paul Bennewitz at the Albuquerque mutilations conference in 1979, the UFO connection to the cow deaths had been firmly established. Then, on 25 May 1980, *A Strange Harvest,* a documentary about the mutilation wave in Colorado, written and produced by Linda Moulton Howe, beamed the ET explanation into the national consciousness. It included a hypnotic regression session with an alien abductee by University of Colorado psychologist and UFO researcher Leo Sprinkle. It's no surprise, then, that when Sprinkle and Bennewitz hypnotized Myrna Hansen, who had come to Bennewitz for help following her apparent abduction in early May of that year, she 'witnessed' the mutilation of a calf – and mutilated humans – inside the underground base into which her captors had taken her.

After the Albuquerque conference, Bennewitz and Valdez maintained a correspondence, eventually undertaking joint missions to look for clues to the mutilation mystery in the Dulce area. Gradually, Dulce began to take on greater prominence in Bennewitz's ET mythology and, by mid-1981, he had come to believe that the aliens were flying abduction and mutilation missions from a base deep inside the Archuleta Mesa. Naturally he told Richard Doty and Bill Moore all about it. AFOSI's work up to this point had been to fuel Bennewitz's delusions with the false documents being fed to him by Moore and the 'ET' messages reaching him via his computer. Doty's role had been to court Bennewitz as a friend and to provide him with gentle encouragement and positive reinforcement that his UFO and ET research was on the right track. Now things took a more theatrical turn.

Sensing an opportunity to shift Bennewitz's attention away from Kirtland, AFOSI began 'set-dressing' the Archuleta Mesa to look more like the underground base that the physicist believed it to be. At night, old military equipment was hauled up the long winding tracks to the top of the Mesa. Shacks, broken-down vehicles and air vents were tactically arranged to give the impression of an active location, while patches of scrub were cleared to look like landing pads for helicopters and, perhaps, UFOs. Doty claims that his team also set up a system for projecting lights on to the clouds above, generating new UFO reports to keep Bennewitz, Valdez and others coming back for more.

An underground base also needs staff, so Kirtland's Special Forces unit was sent out to the area to look busy. AFOSI also contacted Fort Carson Army Base on the Colorado side of the Mesa and suggested that they use it for training exercises. According to Doty, AFOSI even subsidized these Army exercises, explaining that their manoeuvres would be part of an anti-Soviet counter-intelligence operation; which in a strange way, they were. On one occasion Gabe Valdez and a TV crew were shooting a news segment about local UFO sightings with Bennewitz when a Blackhawk helicopter buzzed their own chopper. Panicked, the news crew landed, followed by the Blackhawk. Valdez angrily confronted the black-clad Army men on board, pointing out that they were within his jurisdiction as Highway Patrolman. Before being warned off, Valdez got a close look at one of the soldiers' patches and noted that they were from Fort Carson's elite Delta Force unit.

Bennewitz, himself a skilled pilot, would regularly fly over the Mesa looking for entrances to the alien base, and on at least three occasions was given guided tours by Rick Doty and Colonel Edwards, Kirtland's security chief, who showed him the vents and other paraphernalia that had been placed there by AFOSI. Encouraged by AFOSI, Bennewitz distributed regular reports about the base to the UFO community, incorporating his own photographs of blurry 'UFOs' and indistinct surface features on the Mesa. These gave rise to an elaborate mythology about Dulce's alien base that would incorporate detailed drawings of its interior and dramatic accounts of showdowns between the US military and the base's ET inhabitants.[5]

It's possible that in late 1985 Bennewitz really did stumble on to something unusual. On one of his overflights, armed as usual with his camera, he spotted what he described as a crashed, delta-wing aircraft in one of the more inaccessible areas of the Mesa. Bennewitz immediately reported the sighting to his wide list of contacts, including Gabe Valdez, Bill Moore and New Mexico Senator Pete Domenici. He took numerous photographs of the downed craft, handing some over to the US Air Force while others were quite likely stolen from him – a regular occurrence, according to his letters from the period. All that survives now are his drawings of the crashed plane and photographs of the site after it was cleared away. In his letters, Bennewitz claims that the wreckage was a nuclear-powered test aircraft being flown in secret by the Air Force, shot down by the aliens to remind the government who was in control of the area. The crash site was allegedly saturated with radiation from the downed craft's fuel cells, something Paul was deeply concerned about.

In early November of that year, Gabe Valdez, Bennewitz, a Jicarilla tribesman and a government scientist on loan from Senator Domenici all made the arduous journey to the site, equipped with a Geiger counter. Although they detected no radiation, they did find signs of a crash – broken trees, a gouge in the earth and, according to Valdez, a government-issue ballpoint pen.

Was it a crashed Stealth plane that Bennewitz spotted? It's quite possible. Operational F-117A Stealth Fighters were being flown from at least 1981, though they remained highly secret until 1988. Development on the huge B-2 Stealth Bomber, which Bennewitz's drawings more closely resemble, began in 1981 and prototypes may have been flying by 1985. If one of their aircraft really had crashed on the Mesa then the Air Force may have concocted the radiation story to keep Paul and other curiosity seekers away while a clean-up took place. Although Doty was off the case by this time, AFOSI were still working Bennewitz's number; the irony of the pixie-led physicist finding a downed Air Force craft in an area he was scouring for their own aliens must have set a few eyeballs rolling at AFOSI headquarters.

THE ROAD TO DULCE

All this was more than twenty years ago, but UFOs are still occasionally reported from the Dulce area, and the place retains an unshakeable aura of mystery. Are there any traces of its bizarre past left to be discovered by contemporary curiosity seekers?

Gabe Valdez told John and me that he knew of access tunnels to an old mine on the Mesa that might once have housed a mutilation lab and, to our delight, offered to take us and Greg Bishop up there for a look. However, by the time we were able to make the trip, the Jicarilla tribe had announced a ban on TV crews filming in the area, in part because Gabe had led a Japanese film crew on to the Mesa without permission. Disappointed that we wouldn't make it to the secret labs, we decided to visit Dulce anyway.

While John grabbed location shots up in the mountains, Greg and I approached the Mesa, concocting a cover story in case we were questioned. Greg is a keen paraglider, and Archuleta Mesa is a perfect place from which to throw yourself with a small motor and a parachute strapped to your back. With Greg's chute in the boot we headed on to a small, winding road that leads to the Mesa's base, and then up. It's about three miles to the bottom of the Mesa; we passed a few cows, fully intact, and lightning-blasted trees. Greg and I could feel the Mesa's pull, the frisson of the unknown emanating from its peaks like a radio signal – for us this was a site of pilgrimage. As we got nearer, bright-red signs began to appear at the side of the road, warning us that we were on private Jicarilla tribal land: entry without a permit would lead to arrest and a large fine. Not wanting to spend the night in a Dulce jail cell we decided to turn back and explore another part of town.

In a bleak concreted-over meeting area not far from the Best Western hotel we ran into two drunk Indians, Humphrey and Sherman, both probably in their forties. Alcoholism is a critical problem on Indian reservations; unemployment is high and although Indian residents are given a not-ungenerous subsidy to live at Dulce, too many still succumb to the bottle.

The more drunk of the two men, Humphrey, was wild-eyed and long-haired. Sherm looked a lot tidier and a lot sadder. They were half-brothers, they told us. Like a lot of Indians, Sherm was moved between reservations at a young age, though he felt that Dulce was now his home. While those of us educated by Hollywood Westerns think of the Indians as desert-dwellers, Sherm told us that back in the nineteenth century his tribe, the Apache, originated in verdant Canada and were pushed down into the desert south. And before Canada, they had come from the centre of the Earth.

Sherm admitted to being an alcoholic, but he used to be a tribal policeman and knew Gabe Valdez, who he told us was a good man. He asked us what we were doing in Dulce, so we told him about Greg's paragliding and asked about launching off the Mesa. Sherm's face turned serious:

'No, you can't go up there. That's the danger zone. It's not safe.'

Our ears pricked up. Was he going to warn us about the UFOs or the cattle mutilators?

'It's very dangerous up there, man. The air currents will drag you right back against the Mesa wall, maybe even kill you.'

Hmm. We tried another tack. Did Sherm ever go up there?

'Oh yes. I go up there a lot, man. I like to take my dirt bike up there.'

'So do you see anything unusual? You know, anything that shouldn't be there?'

'Oh sure man, sure. You find animal horns up there sometimes, cattle. You can sell 'em!'

Humphrey staggered our way, eyes rolling, arms waving.

'Yeah I go up there . . . and I go cattle chasing . . . yeah . . . but you don't want to go up on that Mesa, oh no man. Bad things can happen up there. You know . . . people who go up there, they drink too much . . . and then they fall off . . .'

'Hey, you wanna see some secret Indian magic?'

'Sure.'

Humphrey waved his arms towards dark storm clouds gathering over the mountains a few miles to the east. He chanted and gesticulated, jabbing his fingers first at the sky, and then at the ground, part-whispering, part-gibbering something unintelligible to me and, I suspect, even to someone who spoke Apache. Then he looked at me, momentarily sober.

'I will show you Nature!'

Laughing maniacally, Humphrey raised his arms one last time. I was about to laugh back when two lightning strikes crashed loudly on to a nearby hilltop.

I was impressed and told him as much. He leaned conspiratorially into my ear and whispered: 'Don't tell anyone what you saw.' Then he shook my hand and smiled, happy to have put on a good show.

Safely tucked away in the Best Western hotel that night, I had a strange dream. John, Greg and I were on foot in the Dulce landscape, its wide open plains enclosed by steep ragged rocks. A nineteenth-century wagon train hauled itself slowly, creakily in our direction, receding some way into the landscape. At its front was a grizzled, weather-beaten man, his tanned, ageless face merging with the dry leather of his Stetson. The man nodded and smiled, touching the brim of his hat. I smiled back and looked around at

John and Greg to nod my approval at this pleasingly anachronistic vision.

When I turned back around the man, and the wagon train, were gone.

ELEVEN
OVER THE EDGE

'The greatest derangement of the mind is to believe in something because one wishes it to be so'

Louis Pasteur

Rick Doty liked John and me, and we liked him. Over the course of our week at Laughlin he became our almost constant companion, a state of affairs that occasionally caused us anxiety. One day early on in our acquaintance, the three of us decided to pop over the road to a small mall where we could pick up a wireless Internet connection on our laptops. As we crossed the busy main street outside the hotel, really more of a motorway, I was struck by a dizzying wave of paranoia. Rick had by this point already told us about his computer hacking skills, his meetings with DIA colleagues and their interest in some of the UFO conference attendees, especially 'foreign nationals', as Rick always referred to them.

Suddenly it hit me. By going online in the vicinity of Rick and his computer, were we not just opening up our own machines, packed with information about us and our film, to infiltration by the Intelligence machine? As small black stars danced in front of my eyes, I realized that it was too late to back away. I would have to ride the wave and see this through.

Inside the mall's white-tiled, air-conditioned, musak-ventilated interior, we all sat at the same table, John and I on one side, Rick on the other, the backs of our laptop screens facing each other. There was a passcode set-up on the wireless transmitter and the cafe that provided the service was closed. We joked that Rick ought to be able to hack into the system and get us online, a statement that he didn't argue with. After about a minute, however, it became clear that this wasn't going to happen, so I turned around and got the password from a laptop user on another table. We chalked that one down as a triumph of HUMINT (human intelligence) over ELINT (electronic intelligence) and for some reason I never felt anxious around Rick again.

Which isn't to say that he didn't do some strange things. One afternoon, Rick approached me with his laptop while I sat working at my own in the conference foyer.

'Hi Mark!' There was a furtive air to Rick's voice that instantly set me on edge. 'I have something I want to show you.'

He put the laptop down on the table and opened the screen to reveal a series of photographs and drawings of extraterrestrials, all in the 'grey' variety – grey skin, bald bulbous heads, large black almond-shaped eyes, slits for mouths, holes for noses and pencil-thin necks.

'What do you think of these?' he asked.

I'd seen almost all these photos before and said as much to Rick. I also told him that I thought they were all fakes; some were models, some were special-effects creations from films. A couple were very well done.

'Why are you showing me these?' I asked, making no attempt to hide my suspicion.

'One of them is real. Which one do you think it is?'

'I think they're all fake, I've told you.'

'I've seen a living Eben and one of these is a photograph of it.'

He pointed out one of the creatures, a side-on profile of a grey with a longer-than-usual face and a noble, determined look, at least inasmuch as you could read emotion into its alien visage. Its arms were down by its sides and the whole image was cut off above the elbow. I expressed my doubts.

'This one is real,' insisted Rick. 'We called it EBE 2. It lived as a guest of the US government from 1964 until 1984. I saw it being interviewed at Los Alamos.' Rick described this same incident in *Exempt from Disclosure*. On 5 March 1983, he was led by an unnamed source to a vault deep beneath Los Alamos National Labs. Here they entered a room containing 'two tables, several chairs and recording equipment. I sat near the door.' An Air Force colonel entered and told Doty to remain silent during the interview. Rick asked who the interview subject was and was told 'a guest from another planet! . . . About five minutes later, in walks a 4′9″ non-human looking-creature [sic]. It was dressed in a tight fitting cream colored suit. It had no hair and was identified to me as EBE 2.'

The Colonel and two other unknown civilians asked EBE 2 questions, mostly about his home planet and the atmosphere there. 'The interesting part of this interview was that I didn't hear any questions being asked by the three humans sitting across from EBE 2.' Doty heard only the ET's replies, which were in 'perfect English, but sounded like a machine-generated voice'.[1] EBE 2 told them that it liked the climate in New Mexico because it reminded it of home.

As Rick and I looked at the picture of EBE 2, John appeared. I pointed out EBE 2 and repeated what Rick had told me.

'That's a bust,' said John flatly, 'a model. A ufologist I know has it on his mantelpiece.'

A hint of frustration crept into Rick's voice: 'If that's a bust then it's taken from a photograph, because I know that it's real!'

'It's a bust,' insisted John, before wandering off again.

When Rick wasn't with us, he was with Bill Ryan. Rick was reluctant to talk about the Serpo story with us, but the subject always came up when Bill was around. We'd not had a lot of opportunity to speak to Bill during the conference, where he was very much the centre of attention. Almost every time we saw him he was being interviewed by journalists or quizzed by ufologists.

On the Thursday afternoon, John and I were sitting with Rick in the Aviary Lounge waiting to see Bill. A man in black asked to sit with us. We all nodded our consent. He was well-built but somehow effeminate. If you asked me I'd say that his black polo neck was fitted a little too tight. Within moments of sitting down, the man proceeded to tell us about his UFO experiences in a thirty-minute monologue. He spoke as if he were hypnotized or dreaming, in a fey, lilting voice, telling us he was a single parent and that he had seen a flying saucer outside Phoenix, Arizona several years ago. Soon after that, he had met two strange men with flashing eyes. One had eyes that shone red, the other gold. The meeting with the first man had been a largely positive experience and our speaker had taken his phone number. The second was traumatic and left him feeling as if he'd 'been penetrated without being asked for permission'.

Our narrator forgot about his experiences for two months, then found the first man's phone number and called it. The phone was answered by the man's wife: 'He's out,' she said, 'you'll have to excuse me, I have some enchiladas in the oven.' As she hung up the phone the memories came flooding back. All he could remember were the men's eyes, 'spinning and flashing like fireworks'.

Bill appeared as the man was finishing his story. He looked agitated and was clutching a manila envelope.

'Guys, I have something to share with you.'

We excused ourselves from the man in black and sat at another table.

'I was just given this envelope at the lobby. It was left there for me.'

'B. Ryan' was scrawled on the envelope with a thick green marker pen.

'Let me take a look,' said Rick. 'I'm a seals and flaps man.'

Bill opened the envelope himself.

Inside were three sheets of word-processed text, printed on a banded background – the printer must have been running out of ink. It was a new section from the Serpo astronauts' journal, describing their difficulties communicating with the Ebens after being taken to their planet. With it was a page of small, hand-written, squiggly glyphs, forming a wobbly grid of sixteen lines. Was it an alien script? We rotated the page a few times to see which way the squiggles were supposed to flow. We also looked for repeating characters, though there was no chance of our making any sense of it. Bill put the papers back into the envelope.

'Sorry guys, I've got to go. I've arranged to watch UFO videos with some of the conference organizers.'

We arranged to meet Bill in the hotel foyer after dinner. Tomorrow was his big day, time for his presentation. As he bounded off to his screening, I couldn't help but wonder whether or not Bill was the captain of his own ship as he charted the murky waters of alien reality. Within days of his arrival at Laughlin he was being welcomed with open arms by the American UFO community, but I couldn't shake off the possibility that he had been manipulated and conditioned, even if I didn't know who by. My biggest fear, and one that both John and Greg shared, was that he was on his way to becoming another Paul Bennewitz.

PROJECT BETA

Circulated to the UFO community in late 1981, *Project Beta: Summary and Report of Status (With Suggested Guidelines)* is a remarkable and, in hindsight, tragic documentation of Paul Bennewitz's magnificent obsession. In the kind of detail that one would get only from an engineer of Bennewitz's brilliance, it summarizes the data that he had amassed and built into his labyrinthine fantasy. The tragedy is that, unlike other well-known entries in the literature of mental illness, we know that this one was as much a result of the disinformation being fed to Bennewitz via AFOSI, as of his own imagining. It also reveals the extraordinary extent to which AFOSI's perception management operation went to provide Bennewitz with false information.

The twenty-five-page report opens with a summary of the information gathered by 'Investigator – Physicist – Paul F. Bennewitz'.

> Detection and disassembly of alien communication and video channels – both local, earth, and near space. Constant reception of video from alien ship and underground base viewscreen; Typical alien, humanoid and at times apparent Homo Sapiens.
>
> Established constant direct communications with the Alien using a computer and a form of Hex Decimal Code with Graphics and print-out. Through the alien communication loop, the true underground base location was divulged by the alien and precisely pin-pointed. Subsequent aerial and ground photographs revealed landing pylons, ships on the ground – entrances, beam weapons and apparent launch ports – along with aliens on the ground in electrostatically supported vehicles; charging beam weapons also apparently electrostatic.

Bennewitz outlines what he has learned about alien psychology: 'The alien is devious, employs deception, has no intent of any apparent peace-making process and obviously does not adhere to any prior-arranged agreement.'

While he was referring to the extraterrestrials that he believed were inside Archuleta Mesa, he might just as well have been talking about his 'friends' at Kirtland. At times he seemed to be drawing attention to inconsistencies in the information being fed to him: 'In truth they tend to lie, however their memory for lying is not long and direct comparative computer printout analysis reveals this fact. Therefore much "drops through the crack" so to speak; and from this comes the apparent truth.'

Later on he says of the aliens:

Mark Pilkington and John Lundberg in front of the Manzano Weapons Storage Area at Kirtland Air Force Base. (© *Greg Bishop*)

Four Hills, Albuquerque, home of Paul Bennewitz.

Bill Ryan at the 2006 Laughlin UFO conference.

Richard Doty.

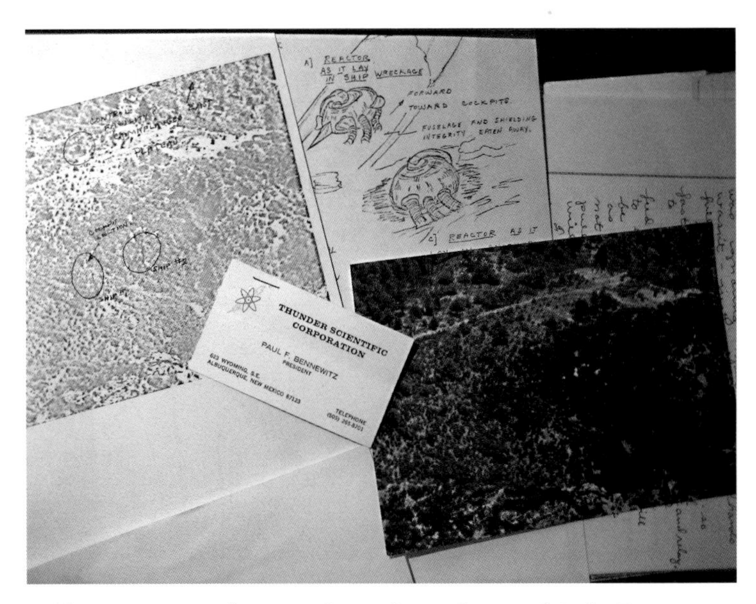

Paul Bennewitz ephemera from the archives of Bob Emenegger: Thunder Scientific business card; drawing of crashed alien craft; photographs of Archuleta Mesa.

Archuleta Mesa at Dulce, New Mexico: home to an underground alien base?

Novel Parachute Plane Is Built to Land in the Back Yard

Recently tested in Chicago, saucer plane made 135 m.p.h. and landed like a parachute.

FIRST cousin to the autogiro, a new circular-wing airplane recently tested in Chicago is so simple in operation that one who has never been off the ground can learn to fly it in thirty minutes, according to the inventor.

Instead of the conventional wing structure, the new plane has a huge saucer-like disc trussed above the fuselage. At the rear of the wing are two ailerons which enable the plane to land at low speeds.

A small 110-h.p. Warner motor develops a speed of 135 miles per hour. The ship climbs at an angle of 45 degrees and lands at a speed of 25 miles per hour, coming to a halt within a few feet.

The plane's peculiar fifteen-foot wing is attached to a conventional fuselage by braces like those of the usual high wing monoplane. The ship carries two passengers and can be housed in a hangar not much larger than the ordinary garage.

The invention of Steven P. Nemeth, former aeronautics instructor at McCook Field, the plane is virtually stall-proof, foolproof and can land on any kind of field.

A 1934 flying saucer depicted in *Modern Mechanix* magazine (*via blog.modernmechanix.com*).

The Vought-Zimmerman V-173 in 1941, an early version of the XF5U-1 Flying Flapjack. (© *NASA/courtesy of nasaimages.org*)

One of a series of communications satellites launched by NASA in the 1960s. Each spent several years in Low Earth orbit and would have looked to the naked eye like a bright, moving star. (© *NASA/courtesy of nasaimages.org*)

VZ-9 Avrocar, all that's left of John Frost's Avro Canada flying saucer project. (© *National Museum of the US Air Force*)

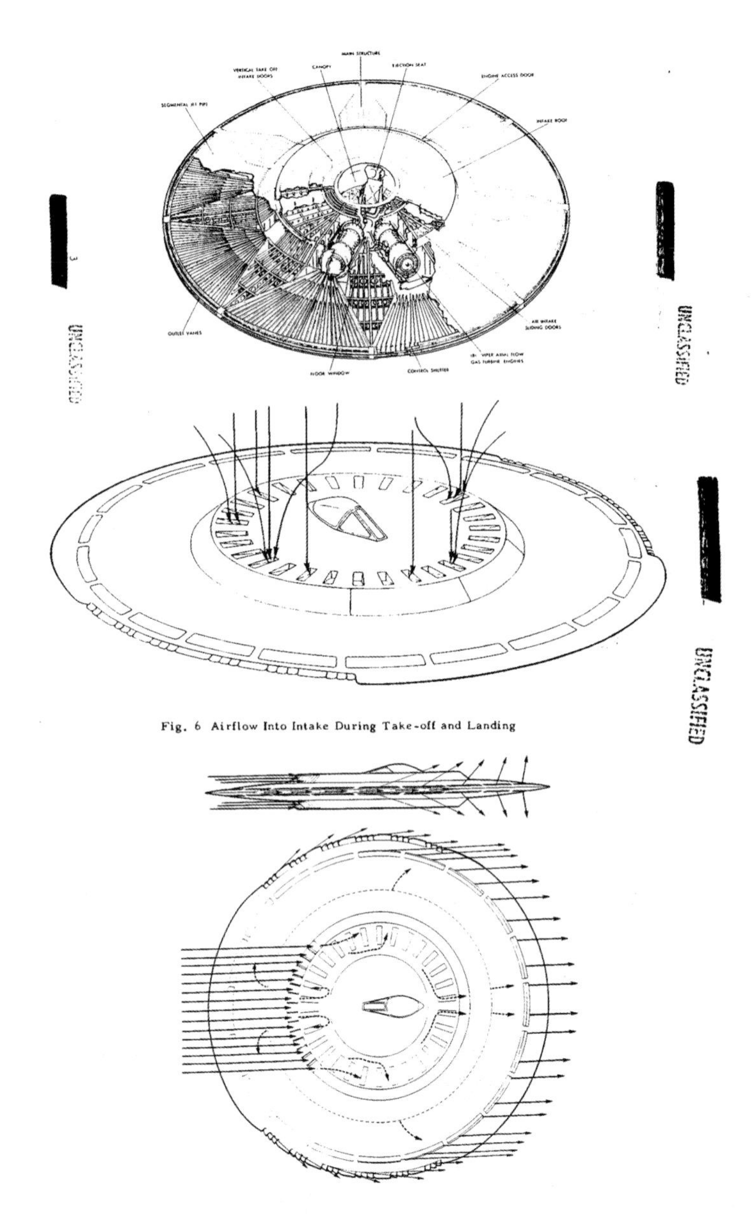

Fig. 6 Airflow Into Intake During Take-off and Landing

Project Silverbug prototype designs.

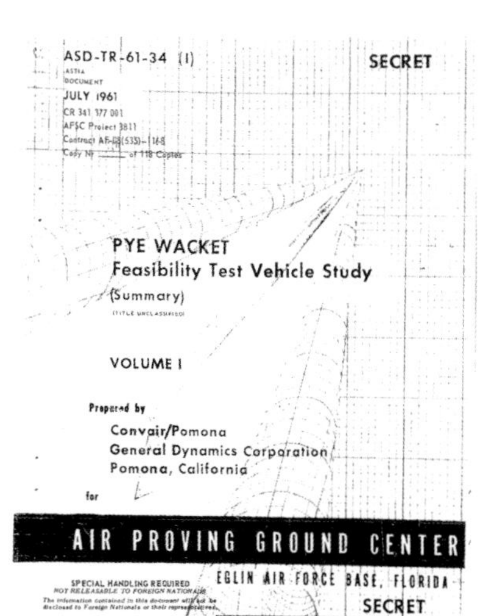
ASD-TR-61-34 (I)
ASTIA DOCUMENT

SECRET

JULY 1961
CR 341 377 001
AFSC Project 3811
Contract AF-08(635)-1168
Copy Nr ______ of 118 Copies

PYE WACKET
Feasibility Test Vehicle Study
(Summary)
(TITLE UNCLASSIFIED)

VOLUME I

Prepared by

Convair/Pomona
General Dynamics Corporation
Pomona, California

for

AIR PROVING GROUND CENTER
EGLIN AIR FORCE BASE, FLORIDA

SPECIAL HANDLING REQUIRED
NOT RELEASABLE TO FOREIGN NATIONALS
The information contained in this document will not be disclosed to Foreign Nationals or their representatives.

SECRET

DOWNGRADED AT 3 YEAR INTERVALS;
DECLASSIFIED AFTER 12 YEARS.
DOD DIR 5200.10

Pye Wacket Feasability Test Vehicle Study, 1961.

SECRET

SECRET

Ryan AQM-34L (Firebee II) reconnaissance drone, used for photographic missions in Vietnam. (© *National Museum of the US Air Force*)

Next generation flying saucer: the Northrop Grumman X-47B UAV. (© *Northrop Grumman*)

> They are not to be trusted. It is suspected if one was considered a 'friend' and if one were to call upon that 'friend' in time of dire physical threat, the 'friend' would quickly side with the other side . . . They cannot under any circumstances be trusted . . . They are totally deceptive and death oriented and have no moral respect for human or human life . . . No agreement signed by both parties will ever be adhered to nor recognized and respected by the alien, though they might attempt to make us believe otherwise.

It's hard not to read these statements as brief moments of unconscious clarity about his situation that, thanks to the effectiveness of the campaign against him, were directed at the shadows on the wall, rather than the hands casting them.

Bennewitz warns his readers of the aliens' ruthless ambitions for domination of the human race, detailing the mind-control implant technologies with which they create human slaves, their weaponry and spacecraft, and the base at Dulce. The report concludes that the only way to neutralize this threat to humanity is by force and outlines an elaborate plan to cut off the water supply to the base at Dulce. Bennewitz also describes a specialized beam weapon that he has developed to counteract the ETs and their craft, something that must have greatly interested his military handlers.

In a chilling final statement he declares:

> The key to overall success is – they totally respect *force* . . . . As Americans, in this particular instance, we must realize that we in this case cannot rely upon our inherent moral principles to provide the answer. Negotiation is out. This particular group can only be dealt with no differently than one must deal with a mad dog. That method they understand . . . Therefore, in eliminating this threat, we most certainly cannot be called the 'aggressor', because we have literally been invaded.

One can only guess at the mindset of those at Kirtland who, faced with such clear signs of mental instability, decided to take their operation to another level rather than attempt to defuse a situation that was already out of hand. And yet AFOSI's campaign against Bennewitz continued until at least 1984, and his mental state continued to decline.

When Rick Doty spoke about Paul Bennewitz, he spoke of him as a friend, and as 'a wonderful human being'. He told us that he felt terrible about what happened to him – and I believed him. Doty's active involvement in the Bennewitz affair ended in 1984, when he was transferred to Germany for two years; Bill Moore's AFOSI work ended the following year. It's a testament to the relationships that the two men built with Bennewitz that they both stayed in contact with him long after the AFOSI operation was over. But by then it was too late to help him.

By the time Doty returned to Kirtland in 1986, Bennewitz's mental health had dramatically deteriorated; Doty tried to encourage him to drop his UFO research for the sake of his family, his business and his health, all of which were suffering. He even told Bennewitz that the UFO information he'd been receiving was the work of AFOSI, but Bennewitz wasn't having any of it. Not only was he convinced of the ET invasion of New Mexico, but by now other ufologists were also listening to him; he had an audience. Together they could perhaps save the world from an alien invasion, but they couldn't save Bennewitz from himself.

By 1987, when Bill Moore last visited Bennewitz, he was barely sleeping or eating. He was chain smoking (Moore told Greg Bishop that he once counted twenty-eight cigarettes in forty-five minutes) and he was having trouble stringing sentences together. He was intensely paranoid, installing extra locks on doors and windows and hiding guns and knives around the house. He told Moore and Doty that the aliens were climbing into his bedroom at night and injecting him with drugs that made him do strange things. Bennewitz would sometimes wake up behind the wheel of his car, having driven out into the middle of the desert. According to Greg, both Doty and Moore independently told him that they had noticed what looked like needle marks dotting his right arm; he wondered whether one or other government agency was injecting Bennewitz with drugs, then driving him out into the desert to pump him full of more alien horrors and absurdities. Doty thought Bennewitz was either harming or injecting himself, though he also claims to have seen ladder marks outside a first-floor window of the Bennewitz home, right where Paul claimed the aliens were entering the house.

Things finally came to a head in August 1988. Bennewitz, aged sixty-one, was barely functioning. His business, Thunder Scientific, was being run by his two adult sons. Back at home, he was accusing his wife Cindy of being in the control of the extraterrestrials. The final straw came when he barricaded himself into their home with sandbags.

It couldn't go on. Paul's family had him sequestered at the mental health facility of the Presbyterian Anna Kaseman hospital in Albuquerque, where he was held under observation for a month. When Doty went to visit his old friend, Bennewitz didn't even recognize him.

So how did it come to this?

According to Doty, the AFOSI operation, which was concerned only with the security of Kirtland AFB, ended in the mid-1980s. However, the National Security Agency (NSA) – who were beaming false signals at him from the house opposite to stop him monitoring their own transmissions from Kirtland – continued their operations for another few years. It's unclear why.

Were the NSA really involved? The faked government UFO memos generated by AFOSI and given to Moore and Bennewitz name NASA as the source of much of the government UFO cover-up. Doty now states that NASA was really an allusion to the NSA, but there were good reasons why the US Air Force may have wanted to drop NASA in the shit. The two organizations had enjoyed a bitter on-off inter-service rivalry dating back to NASA's origins in the late 1950s: the US Air Force had always had one eye on Outer Space, so to see so much aerospace funding going to NASA – a civilian organization – was galling for Air Force hawks. We might also imagine that the Air Force was unhappy at being landed with the UFO hot potato – didn't space issues fall under NASA's umbrella?

UFOs have always been a pubic relations nightmare for the US Air Force, one that began in 1947 and continued until their closure of Project Blue Book in 1969. From this date on, the Air Force replied to any UFO-related enquiries with a form letter stating that they had ceased involvement in the subject. This was a lie, of course. The activities of Falcon, Doty and AFOSI make it clear that even if they had stopped investigating, or instigating, UFO cases – which is unlikely, given the Air Force's custodianship of the skies – they certainly hadn't stopped thinking about them, or monitoring UFO groups.

And the UFOs didn't stop coming. Following the 1975 ICBM sightings in Montana came the release of Steven Spielberg's *Close Encounters of the Third Kind*, on 16 November 1977. An immediate sensation, the film spearheaded a worldwide resurgence of interest in the subject, including a failed attempt by Eric Gairy, President of Grenada, to set up a United Nations UFO research agency.[2] The number of sightings shot up all over the world as a whole new generation of UFO enthusiasts was born, all wanting to know the truth about what was flying in our skies. Responding to growing public demand, President Jimmy Carter, who had spoken openly about his own sighting while on the campaign trail, suggested that NASA should lead a new UFO study.

But NASA didn't want to play ball: 'We are not anxious to do it because we're not sure what we can do,' a NASA spokesman told the Associated Press. 'There is no measurable UFO evidence such as a piece of metal, flesh or cloth. We don't even have any radio signals. A photograph is not a measurement . . . Give me one little green man, not a theory or memory of one, and we can have a multimillion-dollar program.'[3]

Either way the US Air Force was going to lose out: if NASA began an investigation they might want to start prying into the US Air Force's past relationship with the UFO mystery, which would be time-consuming, expensive and not a little embarrassing. If they didn't conduct the investigation then the Air Force would be the next organization to be expected to do so – and they weren't going to go through all that again, barely a decade after they'd successfully divested themselves of the problem.

In the end neither NASA nor the US Air Force were forced to conduct a new UFO investigation, but it was a close call and no doubt stirred up more bad blood between the organizations. So when AFOSI identified NASA as the source of the government's UFO clandestine project in the faked Aquarius documents circulated in 1981, they were both shifting attention away from themselves and taking a pot shot at their rivals. As a result NASA would have got a small taste of the madness and grinding public relations nightmare that swarms of UFO inquiries represented, and that the Air Force had endured for more than twenty years.

AFOSI also knew that Bill Moore was inclined to take their NASA bait. In their book *The Roswell Incident*, Moore and Berlitz list a number of astronaut UFO incidents allegedly covered up by the space agency. Most spectacular is their claim that the landing site of the historic Apollo 11 mission had to be relocated at the last minute because the initial location was 'crawling' with other 'spacecraft'. The book included an alleged transcript of a terrified exchange between pilot Buzz Aldrin and Mission Control, an inclusion that led to Aldrin bringing a lawsuit against the two authors.

But the real culprits in the Bennewitz case may, as Doty insists, have just been a capital A away from the space agency: in the early 1990s after NASA switchboard operators had no doubt grown very tired of taking phone calls from inquisitive ufologists, Bill Moore revealed that the original Aquarius document had indeed read NSA (National Security Agency), not NASA – an alteration very likely made with stifled giggles from the AFOSI specialists involved.

Would the NSA have been interested in Bennewitz? Perhaps, if the signals he was intercepting were theirs and not the Air Force's. Furthermore, the NSA had picked up an intercept from the Soviets that mentioned Bennewitz as a possible useful source of information, though Bennewitz himself could not have been aware of this. This was exactly the sort of scenario that AFOSI and the NSA would have feared most: that Bennewitz's feverish and indiscriminate distribution of UFO information would draw Soviet attention to whatever genuine clandestine research and development programmes were going on at Kirtland.

The real question is why would the US Air Force, the NSA or anybody else want to destroy Paul Bennewitz?

Greg Bishop thinks that the engineer stumbled on to classified aircraft and satellite technology tests at Kirtland; others believe that Bennewitz saw real ET craft being flown in secret. Yet neither answer satisfactorily explains the Air Force's sustained psychological assault on this brilliant but fragile mind. Nor do they explain why the Air Force – or whichever organization was responsible – would test its new toys, especially its most treasured alien toys, within eye- and camera-shot of a housing development when it had several square miles of secluded, open desert in which to do so. Most damning to the 'secret technology' argument is that, if Bennewitz had seen something that he shouldn't have done, the Air Force could simply have asked him to keep quiet about it. As a patriot and a military contractor, he would almost certainly have done so. Should he have refused, the military could have resorted to legal pressures. As Brad Sparks and Barry Greenwood point out,[4] intercepting secret government or military communications would have been in breach of the Communications Act of 1934 and the Espionage Act of 1917. If Bennewitz had been intercepting sensitive NSA transmissions, for example, they could have arrested him, impounded his equipment and shut down his business. But they didn't. Instead they encouraged him in his delusions. Why?

The version of events proposed by Greenwood and Sparks casts AFOSI, and the entire Bennewitz operation, in a more sinister, premeditated light. Kirtland AFOSI probably first identified Bennewitz at Harrison Schmitt's cattle mutilation conference, held on their doorstep in Albuquerque in April 1979. A month later Rick Doty transferred to Kirtland from Ellsworth, where UFO-themed disinformation had been successfully deployed against the *National Enquirer* the previous year. Four months later, in July 1979, Bennewitz began filming lights over the Manzano range at Kirtland, conveniently within eyeshot of his own home, and recording the radio transmissions that he believed were connected to the UFOs.

On 27 January 1980 Bennewitz received his first communication from the aliens themselves. In a 1981 letter to the Assistant Chief-of-Staff for Intelligence at the Air Force, he makes the striking point that none other than Major Ernest Edwards, the Commander of Kirtland's security force and the first person that Bennewitz reported his UFO sightings to, had been present at the first electronic communications session and had 'unofficially provided valuable logistic judgement' to Project Beta. The Air Force had been actively involved with Bennewitz, encouraging him in his delusions, from the very beginning.[5] In July 1980 AFOSI sent Craig Weitzel's letter about UFOs at Kirtland to APRO, who asked Bill Moore to investigate and finally, that September, Moore was contacted by Falcon and Richard Doty and the Bennewitz operation went into second gear.

Considered in this light, the affair takes on a whole new dimension. Bennewitz didn't accidentally stumble on to the UFOs flying over Kirtland, they were a brightly lit lure, flown entirely for his benefit; the Weitzel letter, meanwhile, was expressly intended to draw in a UFO researcher. Both Moore and Bennewitz were baited and trapped; AFOSI planned to use these men as conduits for their UFO disinformation from the very beginning. Their real target was not Bennewitz at all, but the entire UFO community.

THE WAR ON UFOLOGY

AFOSI's stated mission is 'to identify, exploit and neutralize criminal, terrorist and intelligence threats to the Air Force, Department of Defense and U.S. Government'. Much of what AFOSI does comes under the remit of Information Operations, which in *Air Force Policy Directive 10-7* (2006) are divided into three main categories: 'electronic warfare operations (EW Ops), network warfare operations (NW Ops), and influence operations (IFO)'.

It's these 'influence operations' that interest us. They include 'military deception (MILDEC), operations security (OPSEC), psychological operations (PSYOP), counter-intelligence (CI), public affairs operations (PA), and counterpropaganda'. These, the same document informs us, are to be deployed to 'influence, disrupt, or deny adversarial human and automated decision-making while protecting our own. The Air Force may seek to use some of these capabilities to achieve effects similar to those achieved by kinetic weapons.'[6]

Imagine that you are the Air Force. Your goal is nothing less than total air supremacy and to maintain this you need to keep secrets. Lots of them. Operational secrets, tactical secrets and technological secrets in the shape of aircraft, satellites and weapons. Maintaining these secrets, particularly regarding your technology, is absolutely critical. Meanwhile the ufologists are trying to get a peek at your black projects, bombarding you with Freedom of Information Act requests and accusing you of conspiring with DNA-stealing aliens to cover up the truth about UFOs, which you know don't exist. The ufologists' aim is to reveal all the secrets that you are spending many millions of dollars to protect; you are, quite reasonably, going to consider these people a threat, and you are going to want to neutralize them.

By controlling Paul Bennewitz – a noisy proponent of UFO conspiracies who lived right on your doorstep – and Bill Moore, one of the most well-respected members of the UFO community, you have the perfect base from which to launch an assault on this troublesome segment of the population.

It wasn't difficult, once the operation against Paul Bennewitz was under way, Doty repeatedly told us. Bennewitz didn't need much encouragement in his beliefs, just a gentle nudge every now and again. And Bennewitz did his job very well, drawing the attention of much of mainstream ufology towards him and his sensational but simplistic tales of good versus evil, us versus them, alien versus human.

Like the earlier tall tales of Silas Newton, George Adamski and others, Bennewitz's *Project Beta* served to focus and divide the UFO community, creating a wall of noise around the subject that made serious research difficult; many people who might want to take the subject seriously were dissuaded from doing so. It was a masterclass in information warfare, but it came at a terrible price.

When it came to the endgame with Paul Bennewitz, as he teetered at the edge of the abyss, did he throw himself off, or was he pushed?

BILL'S BIG DAY

We had arranged to meet Bill Ryan that evening after he watched the alien interview. Tomorrow would be his big day, the moment of his Serpo presentation. An hour passed. John, Rick and I waited in the foyer. No Bill. We rang his hotel room. There was no answer. John went up and knocked on his door. The lights were off, nobody was home. There was no Bill . . .

The next morning we found him in the foyer. Bill was agitated. He explained that Anonymous, the mysterious source of the Serpo documents, had contacted Don Ware, a senior ufologist on the Laughlin Convention's board of directors, and told him to assess Bill's talk and report back. Bill was now afraid that if anything went wrong he might lose his gig as Serpo's mouthpiece. He could hardly have been surprised. As spokesman for the most significant announcement in human history, he was always going to be on trial. Anonymous was watching.

Despite the nerves, Bill agreed to be interviewed on camera ahead of his talk, but just as we were getting him set up, Rick appeared.

'Hey guys. Something has come up. Can either of you operate a camera? I might have something special for you; one of my DIA contacts wants to screen a movie, one that's never been shown before.'

'What's in it?' asked John.

'An interview with an extraterrestrial.'

'Another one? How many are there? Has this got anything to do with those photographs you were showing us yesterday?'

'No.' Was that irritation in Rick's voice? 'This is different. Nobody has seen this footage and I think my contacts might let you take a camera in and use it for your documentary.'

Bill was waiting for us to interview him and we'd sensed that he had something he wanted to tell us. Now here was Rick offering us the holy grail of UFO history: a real live alien. Why on earth would the DIA want to show it to us? After four days of game playing, our patience was wearing a little thin. 'We're quite busy with Bill here, Rick,' I replied. 'Perhaps our camerawoman could take a smaller camera in and record it while we interview Bill.'

Rick seemed a little taken aback at our rejection of his historic offer, but he agreed to go and ask his people if this might work. Meanwhile we turned our attentions to Bill. He'd been set up with a radio mic and he was hopping about anxiously. Something was eating him, we could tell because he was stuttering, and Bill stuttered only when he was nervous.

'Guys, those new documents that turned up yesterday ... I wanted them to be the centrepiece of my talk today ... I wanted to show people the alien script and get us started on deciphering it.

But I can't. I've been . . . informed . . . that I can't talk about them yet.'

'Informed by who?' I asked.

'I can't tell you that, but I have to obey his wishes.'

'Well it's a shame not to have something new to show the audience here, but you have plenty of other things to talk about.'

'Yes, I suppose so.'

Bill trotted off to the toilet before we sat him down for the interview. We didn't have time to tell him that his radio mic was still on.

Our ensuing interview with Bill was dramatic. The conference had been a psychological mine field for him and he was beginning to show signs of serious strain. Bill had been fêted as ufology's new hero, the man who dared to speak out; every day he had been besieged by well-wishers and interviewers. He was being welcomed into ufology's bosom, snuggling up alongside the great and good at the conference to watch videos of aliens and UFOs. They were holding him close to them, that was for sure, but to what end?

'I feel like my every move is being watched and analysed,' he told us. 'That every conversation is being listened in on.' We said nothing about the radio mic and the toilet trip.

'I just don't know who I can trust any more. To be honest, I don't know how much more of this I can take.'

What about Rick? Could Bill trust him?

'Rick has been a huge help, a guiding light. His own experience in the UFO field has been invaluable to me. He's really steered me through all of this. I don't think I could have done it without him.'

I was about to ask exactly how Rick had been aiding Bill, and whether or not he'd been doing so since Bill first became involved in the Serpo story when, right on cue, Rick strutted into the foyer. He looked like the cat who'd got the cream.

Sensing possible trouble, John steered Rick away from Bill while I quickly wrapped things up. We weren't going to get much further with Rick looming over Bill's shoulder.

We all sat down around a table. Rick's eyes were twinkling.

'So I saw the alien movie. And boy, is it a good one. But they're not going to let you film it. We just had a special screening for the conference organizers, Bob and Terri Brown, a couple of DIA agents and two others I didn't recognize. It was in one of the hotel rooms, all set up for the screening. They were using Panasonic equipment – the government always uses Panasonic.'

We all leant in towards Rick. Was he going to tell us what was in the film or just what brand of popcorn government agents prefer to munch while watching movies together?

'OK. So the footage is old – late 1940s or early 1950s. A car pulls up outside an installation. It looked a lot like Los Alamos to me – I've been there plenty of times. There are vehicles, civilian and military, parked up outside; they look right for the period. A man walks into the building and the camera follows him down a corridor. If you asked me, it was Robert Oppenheimer.'

John asked how the camera was moving. Was it on a track or being carried on a tripod?

'I'd say it was being carried on a tripod, it wobbled a bit as it followed Oppenheimer through two security doors. These only open after his escort has spoken into a squawk box. We reach another door at the end of this corridor and there's an MP, a military policeman, standing guard outside. I could see the name on his badge. It was Douglas. His uniform and his weapon would fit the 1940s or 1950s timeframe. This door opens, and . . . there it is, the Eben. It had some kind of vocalizer around its throat so it sounded a little machine-like when it spoke.'

'What do they talk about?' Bill asked, excited. 'Do they talk about its home planet?'

'They talk about all sorts of things. And, yeah, the Eben talks about its planet. It's forty light-years away, it has two suns, it's an arid, desert landscape.'

'It sounds just like Serpo!' Bill was beaming now. The anxiety of a few minutes ago had vanished.

'Yes, I suppose it does,' replied Rick.

Bill headed off to get ready for his talk; he was on in twenty minutes. There was a new spring in his stride. I couldn't help but feel that we had just heard Rick's idea of a pep talk.

There was a full house for Bill's presentation. The sound dropped out a few times and he perhaps rambled a little bit, but he was telling people exactly what they wanted to hear: alleged documentary evidence of extraterrestrial contact with the US government, backed by inside sources, with hints of much more to come. He kept his audience's attention until the very end.

Except for one table. The men I'd spotted earlier in the week, the steely-eyed flight jacket man and his two burly attendants, were seated impassively at the same table as always. When Bill mentioned the name Paul McGovern – Rick Doty's alleged DIA source who had originally corroborated much of the Serpo story in emails – Ol' Steely-eyes suddenly looked uncomfortable, got up from his table and left the room. Could he have been Paul McGovern?

After the talk, people lined up to ask Bill questions; I hopped out into the foyer and watched the rest of the audience leave. Emerging from the hubbub was Rick, who sidled up to me and asked what I thought of Bill's presentation.

'It could have used some editing,' I said, diplomatically.

Rick agreed. 'I'll give him a C+. He needs to use slides. You should use slides in a talk, like a pilot uses a compass – they should lead you.'

Bill emerged from the speaking hall, almost carried through by a throng of fans, and sat down at a table. As he answered their questions he paid particular attention to an attractive woman in her early forties; she was short with dark features and hair. Bill and the woman talked for some time. Rick and I smiled knowingly at one another.

Once the hordes had subsided John and I got to talk to Bill, who seemed happy that it was all over and had gone down well. He had been almost paralysed with anxiety beforehand, he told us, not helped by the organizers who, just before he went on stage, had told him that if the lights went out he was to hit the deck immediately. They had been warned that there might be trouble during Bill's presentation, and had taken the extra precaution of having a couple of security heavies waiting offstage.

I mentioned Ol' Steely-eyes's sudden exit and pointed him out to Bill as he slid through the foyer. I suggested that he could even be Paul McGovern, but Bill said no, the man had already spoken to him. He was called Terry and was a UFO contactee from Arizona. He went to all the conferences, he had told Bill. I wondered why a contactee would look so mean, and need a couple of big guns positioned on either side of him at each talk. Were they all contactees? Did the three of them go to all the UFO conferences together?

Later that afternoon I was waiting for Rick in the foyer as we had planned to film some shots with him. Uncharacteristically he was almost an hour late, which gave me time to chat to a physicist who was also a practising occultist, an Internet radio host and a journalist writing about the conference for *Hustler*. When Rick finally arrived he seemed to have forgotten our arrangement, but was happy enough to do the shots anyway.

We returned towards the hotel lifts and turned the corner, walking slap bang into Ol' Steely-eyes. He and Rick recoiled from one another like magnets, looking momentarily startled to have run into each other before making a great display of shuffling their feet and looking at the ground. The lift door opened and we squeezed in for a tense two-floor ascent, Rick and Ol' Steely-eyes studiously ignoring each other like a pair of riled tom cats.

When we reached our floor Rick and I exited with a sense of relief and began to cross an outdoor mezzanine.

'So was that Paul McGovern?' I asked, trying not to smile too much.

'How did you work that out?!' Rick looked shocked and delighted.

I laughed and explained that I'd had my eye on the man and his heavies all week, that I'd noticed the nod pass between him and Rick, that he'd left Bill's talk at the mention of McGovern's name.

'Good work – you should have a job at MI6! But you didn't see us watching you and John from the hotel bar while you ate noodles at Panda Express!' Rick chuckled loudly and smiled at me, pleased with himself. I got the strong impression that this was all a game to him. If only I knew what kind of game.

The shooting went well; Rick was a natural on camera. He talked a little about Paul Bennewitz and the need for intelligence agencies to monitor conferences like this one. He also insisted that he had absolutely nothing to do with the Serpo material. He had simply taken an interest as a private citizen.

That evening a film about crop circles was showing in the main hall and John and I decided to watch it. We'd been thinking about the parallels between the work of the crop circle makers, ourselves included, and the disinformation work that Rick had been engaged in. Both groups operated clandestinely and had created new chapters in the UFO story that had taken on lives of their own, spawning books, conferences, Hollywood movies and powerful systems of belief. Both groups had watched from ringside seats as the mythologies surrounding their activities grew ever larger and more complex. Both groups were forced to keep silent about their work then, when they had spoken out, as Bill Moore and some of the circle makers have done, had found themselves denied their authorial roles by those who believe in the myth. In fact the only real difference between us circle makers and the disinformation artists is that they get paid for their work, and we don't. Any organization that did come clean about its role in developing the UFO myth would immediately find itself in a similarly absurd situation. The believers don't want to know the truth, they only want to have their pre-existing beliefs confirmed and elaborated upon.

John and I entered the darkened hall to the familiar sight of flattened wheat in the fields of Wiltshire. And there was Rick, sitting alone at a table, where we joined him. Every now and again John and I struggled to suppress chuckles as one crop circle expert or another chucked out statements about abnormal radiation readings or genetically altered crops. A few of our own team's designs glided majestically past. This, I thought, must be what an AFOSI agent goes through every time they flick through *UFO Magazine*.

Our amusement was drawing Rick's attention. John leant over and told him that some of the formations were his handiwork. Rick looked amazed. 'How do you do that?!' he screamed in a whisper. We were more surprised than he was; surely he knew that people made the crop circles? And surely he'd checked out John's Circlemakers web site before meeting us?

Footage of two balls of light creating a huge crop circle played on the screen in a film known in the field as the Oliver's Castle footage. 'I made that,' said John. And he had, in 1996, along with two other circle makers and a digital special-effects technician. Rick was unmistakably impressed, and the hushed gasps from the audience reminded us that he was right to be.

After the film Rick tried to play down his surprise. 'Hey, I knew about all that,' he insisted, wiping his glasses on the corner of his shirt. 'I knew that people made crop circles. You didn't burst my bubble! But . . . how do you do it?'

Outside we talked about crop circle making, about UFO hoaxing and about digital trickery; how it's almost impossible these days to determine whether or not a UFO video is faked, and how we've reached a sorry state of affairs whereby the clearer the footage of a strange craft, the less likely it is to be real.

'So,' John asked, 'are you sure that the alien interview footage you saw this morning wasn't some kind of special effect?'

'Oh yeah, that was real.' A sly expression came over Rick's face. He smiled. 'Say, why don't you go and see the film for yourselves.

It was in room 11012; just go knock on the door. They might let you in.'

John and I looked at each other, grinned and headed for the lift, arranging to meet Rick back at the hotel bar.

In the lift we suddenly became nervous. We had only moments to formulate a plan.

'So we knock on the door and say "Hi, is this where the alien film is showing?" We can say that someone approached us in the conference foyer and told us that there was a special screening in here. The worst they can do is tell us to bugger off.'

'Unless they've been watching us all this time,' said John. 'They must have seen us with Rick. They'll know we're with him. They'll know Rick sent us.'

'But what can they do?', I replied, trying to mask my unease. 'We're only going to knock on the door . . .'

The corridor was empty. And long. It narrowed claustrophobically as it stretched away from us. I began to notice the hypnotic repetition of the tiled motif on the carpet, the flicker of the phosphorescent lights overhead. My palms were sweating.

Ahead we spotted a single 'Do Not Disturb' sign hung on one of the doors, the only one on the corridor. Room 11012. Typical. We stepped up to the door and listened. Silence. I took a deep breath and knocked. Nothing. At any moment I expected to see Ol' Steely-eyes and his heavies barrelling down the corridor towards us.

'Knock again,' whispered John.

I did. No reply. I peered through the keyhole. The lights were off. There was nobody home. We considered slipping a note under the door, but instead settled for turning the 'Do Not Disturb' sign around, as a sort of calling card.

We walked hastily to the lifts and wondered what that was all about. Another game? Whose room was that? Was it the DIA's screening room? Or was it Rick's?

Back at the hotel bar we drank bottled beers. At least we had tried.

Our week at the conference was coming to an end, and not a moment too soon. We'd barely talked about anything but UFOs and extraterrestrials for most of that time. The rope that kept us moored to reality was starting to fray around the edges; another week of this and we might easily have found ourselves adrift.

Is that what happens if you spend too long gazing at the UFO's seductive glow? Is that what was happening to Bill Ryan? Is that what happened to Paul Bennewitz?

TWELVE
SEEING IS BELIEVING

'The purpose is . . . the conditioning of billions of human minds, through direct access to their television screens . . . Whoever controls information governs the world . . . The message is no longer obvious; instead it is impressively seductive'
Lofti Maherzi, Algerié Actualité, *13–19 March 1985*

Back at the hotel bar, Rick was telling us stories again. Woody Allen, we heard, was a huge UFO believer and had once hired a private detective to sniff out the truth for him; the guy had spent $15,000 on Allen's credit card and then disappeared. Rick had been to Steven Spielberg's home and seen his F-16 and C-130 flight simulators, bought from the Air Force. Rick had worked as a consultant for the *X Files* and Steven Spielberg's alien abduction series *Taken*. Quite a few of the military's UFO insiders had ended up with these sorts of gigs. Rick told us he'd had a cameo role as a vampire in one *X Files* episode.[1]

I couldn't help but be a little disturbed by the idea that a man who had been part of the programme to destabilize Paul Bennewitz and pollute the UFO research community had then been given access to millions of eager minds – many of them destined to be the UFO researchers of the future – via their televisions. And were any of those consultants still working for the Air Force or the DIA when they extended their myth-making into the popular consciousness?

I asked Rick why, if they know the truth about UFOs and aliens, the government still feels that it has to keep it a secret from the rest of us. It may have been a big deal in the 1950s, when aliens equalled Communists and were something to be afraid of, but post-*Close Encounters* and *ET*, aren't we, the people, primed and ready to meet our intergalactic neighbours? Surely contact would just be another news item, a talking point for a few weeks or months that would inevitably be swallowed up once again by the latest celebrity and sporting news.

And if we did make contact with ETs back in 1947, then what have we got to show for it? What has changed in the world? We're still fighting each other over resources and destroying our environment. Where are the intergalactic policemen like Klaatu? Where are the anomalous leaps in our technological development? Where's our free energy? Our personal flying saucers? Our matter transporters? Where's my sodding jetpack?

'You just said it,' Rick came back at me. 'It's the free energy. The aliens have free energy, and the people in control are afraid of what would happen if the rest of us knew about that. An unlimited source of clean energy would shatter the oil economy and turn the world upside down. It could lead to total chaos: *that's* what they're worried about.'

I wasn't convinced. Even a free energy source, should it ever emerge, will be metered; somebody has to pay for the infrastructure. But yes, I conceded, the change-over scenario might be complicated. Rick got up suddenly and left the table. Had I upset him? A few minutes later he was back clutching beers for us and a club soda for himself. He sat down and smiled at us from across the glass-topped table. The flickering lights of distant slot machines played about his head like flames.

'I like you guys,' he said, still smiling. 'You're smart.'

'Thanks.' We agreed that we liked Rick, too. Something was up.

'We've spent a lot of time together this past week. Did you ever wonder why?'

'Uh, we thought we were just getting on.'

'Well, yes, we are getting on. But I've also been watching you. I had to check if you are who you say you are.'

'How do you mean, on the Internet?'

'Yes, but I also got your fingerprints and had you run up on various databases.'

John and I looked at each other, incredulous.

'Fingerprints?! How the hell do you do that?' John asked. A quiver had crept into his voice.

'Oh, that's easy,' replied Rick, smiling as we squirmed. 'I can run them off anything you've touched. Like a beer bottle.' He picked up one of our beers. 'These are covered in prints. But anyway, don't worry, you're not on any lists, you're clean, you checked out.'

'So what kind of lists?'

'I checked that you're not foreign agents, criminals, anything like that.'

'Well we've made crop circles, that's a crime!' John laughed nervously.

'Don't worry, I'm not going to arrest you for that!'

There was a pause. I sensed the mood lightening.

'OK guys, I've been talking to my friends in the DIA, in the government, and they suggested something to me that you might be interested in.'

Oh shit.

'I know you're working to a very tight budget, and they thought that they might be able to help you out a little bit with that. Perhaps they could give you some money towards your film . . . if you were prepared to take some suggestions about the sort of film that you were making.'

A heaviness descended. John had gone pale. I felt a little bit dizzy. Had Rick put something in the beers? He continued.

'They might give you some material to include in your film – like the movie I saw yesterday – or they might choose to direct the whole project and pay for the entire thing, and for your time on it. They might give you a quarter or a half a million dollars, I don't know . . . Is this something that might interest you?'

I cleared my throat and opened my mouth. John spoke. 'We're really not in it for the money, Rick. If we were, we wouldn't be doing this. We'd be making a film about something else.'

'I appreciate that you're ethical, and I'm glad that you are; it's one of the reasons that I like you. You're not judgemental, you don't take sides, you're just trying to make sense of things. I respect that and I'd be the same in your situation.'

'Well,' I did my best to sound businesslike, 'we'd certainly take a look at their proposal. Perhaps we would suggest some changes, but who knows, they might have some good ideas, or some good footage for us. I suppose it would be like making a film for a big studio, or for an advertising agency. Except we'd be advertising UFOs!'

'Yeah, you could look at it that way. Anyway, I'll leave you guys to think about what we've talked about. I'll be heading back to New Mexico tomorrow. Let's meet before I go and hopefully we can catch up again in Albuquerque next week.'

Once Rick was out of sight, John and I let out a collective sigh of relief and burst out laughing. We could hardly believe it. This was something that we'd always anticipated might happen, though we would never have imagined that it actually *would* happen. And anyway, what *did* just happen? Was that real or was it another one of Rick's little games?

We agreed that we'd be foolish not to at least consider any proposition they passed our way. After all, we could always use it to our advantage and maintain the upper hand. What could be more appropriate for a film about government disinformation and UFOs, than some actual disinformation? If they wanted us to make a pro-ET puff piece, then of course we'd have to tell them where to stick their money – politely, of course. And if we didn't accept their offer, then what could be better than opening our film with a title card: 'During the production of *Mirage Men*, America's Defense Intelligence Agency offered us half a million dollars to produce the film they wanted us to make. We declined their offer.'

As it turned out, Rick's indecent proposal came to nothing. He gave us the name of an intelligence contact who he said would be getting in touch with us to talk business; but after a few weeks' eager waiting, John and I accepted the fact that either the US government wasn't interested in us, or Rick's contact had never existed. Either way, ours would not have been the first UFO film to benefit from the involvement of the American defence establishment.

Back in 1953 the CIA's Robertson Panel had recommended that a 'broad educational program' should be put in place to 'strip the Unidentified Flying Objects of the special status they have been given and the aura of mystery they have unfortunately acquired'. Among the companies named to work on these educational programmes was Walt Disney Inc. and according to one of its lead animators, two years later this is exactly what happened. Ward Kimball was one of Walt Disney's inner circle of animators and designers. He created Jiminy Cricket for *Pinocchio* and the crows in *Dumbo*, and won Oscars for two of his Disney shorts.

In the mid-1950s Kimball wrote and directed three TV specials featuring the German rocket scientist Wernher von Braun: *Man in Space*, *Man and the Moon* and *Mars and Beyond*. Hugely popular, the three 'science factual' films were instrumental in generating popular support for the US space programme and inspiring a generation of children to dream of one day themselves being astronauts. The stern, avuncular von Braun displayed his plans for four-stage rockets, space stations, and atomic-powered spacecraft that would ultimately take man to Mars and beyond. President Eisenhower is supposed to have requested a personal copy of *Man in Space*, as did Leonid Sedov, one of Russia's foremost space scientists.

Ward Kimball was also a keen UFO enthusiast and remained one throughout his life. In 1979 he made an unscheduled appearance at the Mutual UFO Network's annual conference, where he told the audience that in 1955 the US Air Force had approached Walt Disney with the suggestion of making a documentary film about UFOs. The Air Force promised to supply Disney with real UFO footage, and Disney set his animators to work designing alien characters to appear in it. The Air Force never delivered on the UFO footage, leading Disney to cancel the project, although some of the aliens appeared in a fifteen-minute film about UFOs that was never publicly shown.

Kimball believed that the purpose of the Air Force–Disney collaboration was to prepare Americans for the reality of contact with extraterrestrials. It's a rumour that would attach itself to both Robert Wise's Christian UFO allegory *The Day the Earth Stood Still* (1951) and Spielberg's epiphanic *Close Encounters* twenty-five years later. While we can't say for sure what motivated Steven Spielberg or Robert Wise, the probable answer is that they are as influenced by what they read in the newspapers or in UFO books as the rest of us, and their films are a product of this influence.

There is, however, one film that was fully authorized by the Air Force, and its message is clear: UFOs are real.

UFOS: PAST PRESENT AND FUTURE

Bob and Margaret Emenegger are a friendly, lively couple in their late sixties and early seventies; they live in a large, beautiful house filled with English antiques in the verdant farmland of Arkansas. They met while working together at the huge Grey Advertising company in Los Angeles, where Margaret was a designer and Bob eventually became creative director. Margaret is funny, sharp and opinionated; Bob is no less funny but gentler and more laid-back in character.

Although both are retired, they remain active in their community. Bob organizes local music events and plays with a local orchestra. He's a gifted musician and humourist, who's proud to have written the songs and music for the the popular early-1970s television series *Lancelot Link: Secret Chimp*, in which the eponymous chimpanzee plays in a rock band while doubling as a secret agent.

Bob and Margaret believe that extraterrestrials have visited our planet and are probably still here. Margaret also likes to talk about the theories of footballer-turned-conspiracy-theorist David Icke, about questions surrounding the events of 11 September 2001, and about the Indigo Kids, the psychically advanced hybrid offspring of humans and aliens. Bob may not share all of Margaret's views, but he does know something about UFOs. As a film-maker he probably got closer to the truth of the matter than almost any other civilian, and he was invited to do so by the United States Air Force.

In the early 1970s, Emenegger and Allan Sandler, his production partner at Grey, gained a reputation for revitalizing tired corporate brands, like the campaign to re-elect President Richard Nixon, and in 1972 the duo were approached to work their magic for the Pentagon. These were difficult days for the Department of Defense: the Vietnam War had been dragging on for close to a decade and what little trust most Americans had in their government had been reduced to almost nil. The Pentagon needed a boost, both for its own self-esteem, and to encourage people to enlist. Could Sandler and Emenegger help?

It was decided that their film should focus on some of the more exotic, exciting things going on in the military at the time. Emenegger remembers seeing dolphins being trained by the Navy, new developments in atomic fusion, mind-to-computer interfaces, dogs trained in reconnaissance and bomb location. Some of the dogs, Bob recalls, had microchips inserted into their heads – hardly popcorn-munching material.

New technologies such as advanced laser research and 3D holography were also on the menu. Allan Sandler was treated to a particularly impressive holographic demonstration in a screening room with a small stage at one end. The curtains parted and a man walked on to the stage to introduce the Pentagon's new, state-of-the-art holographic projection technology. All of a sudden, a small bird flew out from the wings and landed on the man's shoulder; he smiled and both of them disappeared. They *were* the demonstration.[2]

Dogs, dolphins and lasers were certainly interesting, but they didn't quite cut it. The Grey men wanted something more dramatic, and their man at the Pentagon was going to provide it: UFOs. Sandler and Emenegger were invited to Norton Air Force Base outside Los Angeles, where they met the base's AFOSI head and Paul Shartle, a security officer with AFOSI connections who was also head of film acquisition for the Air Force.

It had been only three years since the Air Force had shut down its UFO investigation, Project Blue Book, so to be suggesting that Grey make a promotional film on the subject made little sense. But the Pentagon was serious and the duo were introduced to two Air Force men at the top: Colonel George Weinbrenner and Colonel William Coleman. Weinbrenner was commander of the Foreign Technology Division (FTD) at Wright Patterson AFB. FTD, now the National Air and Space Intelligence Center, was, and still is, the Air Force's hub for any technology that isn't theirs. If anyone knew about the real UFOs, it ought to be Weinbrenner. Or perhaps Coleman, who had been public liaison officer for the Blue Book in the 1960s and was now Chief Information Officer for the Air Force itself.

In his Blue Book days, Coleman once told American chat show host Merv Griffin that he would believe in UFOs when he could kick the tyres of one. His statement prompted a flood of letters from people pointing out that UFOs didn't need tyres. What he didn't mention at the time was that he had seen a UFO himself, while flying a B-52 over the Alabama/Florida borders in 1955. It was a perfect silver glinting disc that had got so close that he'd had to swerve the giant bomber to avoid a collision. The disc vanished and reappeared about a hundred feet off the ground, two thousand feet below them, creating a shadow and leaving big dust trails.

Hallucinations don't kick up dust. Coleman and his four crew filed an exemplary report and passed it on to Project Blue Book. When Coleman ended up working for Blue Book a decade later, he was surprised to find no record of his report. Interviewed in 1999, Coleman generously put that one down to mismanagement, though UFO hounds had long claimed that Blue Book was a front for another more secret operation. This was finally demonstrated to be true with the release in 1979 of a 1969 memo signalling Blue Book's termination. A key portion reads: 'reports of unidentified flying objects which could affect national security are . . . not part of the Blue Book system'.[3]

Colonels Coleman and Weinbrenner were keen to demonstrate the Air Force's interest in UFOs, so after being offered open access to the Air Force's files, that's what Sandler and Emenegger decided to make their film about. The colonels warned the film-makers of the perils of working with sensitive classified information, then promised them reams of data, photographs and film footage of UFOs both in our skies and in Outer Space. They also hinted at evidence of a close-up UFO sighting involving high-ranking CIA officers and, most dramatically of all, footage of an actual ET spacecraft landing at Holloman Air Force Base as recently as 1971. The film, *UFOs: Past Present and Future*, was going to demonstrate that not only was the US Air Force seriously interested in UFOs, but that it knew what they were and who was inside them.

The colonels would stop at nothing to get this material to the film-makers. Sandler approached NASA to ask for film and photographs of UFOs in space; the space agency gave him their standard rebuff and denied possessing any such material. When Sandler told his US Air Force contacts this they gave him a detailed breakdown of NASA flights on which UFOs had been spotted, the names of the astronauts involved and even the frame numbers of the relevant pieces of footage. Returning to NASA with this new information, Sandler got what he was looking for, though the images themselves contained nothing more dramatic than indistinct blobs and blurs.

Why the Air Force went to such great lengths to provide the film makers with such ambiguous material is just one of the many questions surrounding the affair.

More puzzling still was the story told to the film-makers by Lieutenant Colonel Robert Friend, who contacted them via Coleman. Friend had worked for AFOSI but, between 1958 and 1963, then a major, he had been head of Project Blue Book, which would be reduced to two staff members by the time he left.

In early July 1959, Friend was asked to meet two Naval Intelligence commanders and a huddle of CIA officials at the CIA's National Photographic Interpretation Center (NPIC). Set up in 1954 specifically to analyse photographic data from U-2 overflights, NPIC was hidden on the top floor of a parking garage at Fifth and K Street in Washington DC. The two Navy commanders had visited a woman called Frances Swan in Maine, who claimed to be in telepathic communication with an extraterrestrial called AFFA, the leader of an organization called OEEV that was conducting a research project on Earth known as EUNZA. Sceptical, the Navy men began asking Swan complex technical and astronomical questions which, to their amazement, she answered correctly. AFFA then suggested that he could switch channels and communicate through one of the Navy men, who continued answering tricky questions put to him by his colleague.

When word got back to Washington about these bizarre goings on, the two commanders were summoned to the CIA photographic centre. Here, on 6 July, they repeated the experiment, the same Navy man acting as the conduit for AFFA, who was asked for proof of the aliens' existence. AFFA told the men to go to the window, at which point a flying saucer flew slowly past at a close distance. The stunned men contacted a local radar centre, who reported that they had nothing on their scopes. It was at this point that they called in Robert Friend.

Three days later, Friend witnessed another channelling session, with AFFA once again speaking through the Navy commander.

This time when Friend asked if he could see a flying saucer, he was told that the time was not right. Although disappointed not to have been treated to a flyby, Friend was convinced that the Navy commander's trance was genuine and filed a report to his commanding officer at Wright Patterson.

Through Friend, Bob Emenegger got hold of the CIA's memo about the meetings, apparently written by Arthur Lundahl, NPIC's founder, who had conducted photographic analysis for the CIA's Robertson Panel.[4] The memo confirmed the events as described by Friend and these were featured as evidence of ET contact in Emenegger's film. However, when contacted in 1979, Lundahl denied that a flying saucer had appeared at the NPIC's window on 6 July 1959. He also denied ever having written the memo:

> Never for a fleeting moment did I believe that this Navy officer was in communication with outer space, nor did I see a UFO. The demonstration was not done at our request. The man explained that Mrs. Swan had shown him something called 'automatic writing', and that if asked he would show me . . . He probably chose me because I was a friend . . . Though I believe in intelligent life other than ours, I felt nothing but sympathy and embarrassment on this occasion, for a man who was troubled, who was my friend, and who, if his superiors had learned of this, would undoubtedly have suffered in his career.[5]

So if Lundahl is telling the truth, who prepared the CIA memo given to Emenegger, and why did they alter the details of the story to give credence to the extraterrestrial agenda? Lundahl makes clear that the meetings really took place and that the CIA and Naval Intelligence were involved; Robert Friend really was called in to investigate, but critical details, specifically the appearance of a genuine flying saucer, had been added to the report to make the case more sensational, fuelling the fires of the saucer believers and inducing uncertainties in the unconvinced. This looked like a classic piece of AFOSI disinformation, presented to Emenegger a full decade before the Bennewitz affair.

But the real icing on the cake for the film-makers was to be the Holloman UFO landing that Emenegger was told had taken place early one May morning in 1971. The story, as he heard it, was that three craft, flying discs, had appeared over the base; one of them was wobbling erratically and began a descent. It hovered briefly, a few feet off the ground, before settling there, supported by three legs. By chance, an Air Force film crew in a helicopter had managed to film the descent, while another crew, preparing for a missile launch, had captured the scene from the ground.

In the million-selling book tie-in to his film, Emenegger describes the scene as told to him by Paul Shartle, the Air Force film archivist, who claimed to have witnessed it:

> The commander and two officers, along with two Air Force scientists, arrive and wait apprehensively. A panel slides open on the side of the craft. Stepping forward there are one, then a second, and a third – what appear to be men dressed in tight-fitting jump suits. Perhaps short by our standards, with an odd blue-grey complexion, eyes set far apart. A large pronounced nose. They wear headpieces that resemble rows of a rope-like design.
>
> The commander and the two scientists step forward to greet the visitors. Arrangements are made by some inaudible sort of communication and the group quickly retires to an inner office in the King 1 area.[6]

What happened next, what the aliens talked about, what they had to eat, whether or not they brought gifts, remains unknown.

While preparing to shoot at Holloman, Emenegger asked one of the air traffic controllers about the landing. He remembered seeing an object like a 'flying bathtub' land, but declined to discuss it further. Emenegger was taken out to Building 930 on Mars St, where the aliens and their craft were kept for the duration of their visit, though there was nothing unusual there now. Having shot their scenes at Holloman, Sandler and Emenegger awaited delivery of the footage that would transform their project from a UFO documentary into a historical event. The film never materialized.

A disappointed Colonel Coleman told them that the Pentagon's top brass had decided that the time just wasn't right to open up this particular can of worms; Watergate and the collapse of the Nixon administration had given everyone the jitters. They were advised instead to present the incident as something that *might* happen in the future. As consolation, the film-makers included some footage showing an indistinct object descending on to what may or may not be the runway at Holloman. Although rumours persist that this is a fragment of the genuine landing footage, Emenegger says that it was an experimental craft caught landing at Holloman by his cameraman. As for the alien encounter, Sandler and Emenegger had to make do with using an artist's impression based on Paul Shartle's description.

An irate Emenegger called a meeting with Colonel Weinbrenner at Wright Patterson and asked why the footage had been snatched away from them. Weinbrenner began grumbling loudly: 'That damn MIG 25! Here we're so public with everything we have. But the Soviets have all kinds of things we don't know about. We need to know more about the MIG 25!' He pulled a copy of J. Allen Hyneck's *The UFO Experience*, published the previous year, from the bookshelf and showed Emenegger the signed dedication to Weinbrenner inside. 'It was like a scene from a Kafka story,' remembers Emenegger.

So did the footage ever exist or was it only ever 'dangle', a lure to convince the film-makers that their 'documentary' was going to be something more than just a false advertisement for UFO reality? Paul Shartle insists that he saw the real film, which was bought from a film studio to use as training footage by the Air Force, but describes it, unhelpfully, as 'too real' to be fake.

UFOs: Past Present and Future ends with an interesting coda, in which a group of sociologists survey public beliefs on the subject to find the best way of releasing the truth about ET contact.

One of the sociologists expresses concern that the ETs should not appear too advanced or superior to their new American friends, who might fear that the ETs would treat them like the first settlers treated the Indians. He recommends that the government demonstrates that while the ETs have a lot to offer Americans – 'universal peace, a cure for cancer or solar energy' – Americans also have a lot to offer the ETs: like 'jazz, achievement motivation and Colonel Sanders' chicken.

Another sociologist quotes the psychologist A.C. Elms: 'what we really need is an enemy invader from outer space; then we would unite as one species to drive the invader away and live in peace thereafter', a theme familiar to us from Bernard Newman's *The Flying Saucer*. Most curious of all is the anonymous sociologist who, considering the impact that a film like the alleged Holloman footage might have if shown on television, refers to the hypothetical material as 'Humanoid Organisms Allegedly Extraterrestrial' – HOAEX. We can make of that what we will.

The research is summarized by Leon Festinger, author of the classic UFO cult study, *When Prophecy Fails* (1956). Festinger and two colleagues joined a Chicago UFO cult surrounding housewife Marion Keech, who received extraterrestrial messages from the planet Clarion. When Keech's prophecies of a global deluge on 21 December 1954 failed to take place, rather than leave the group, many of her followers became even more dedicated in their beliefs. Festinger found that if someone believes something to be true, and all the evidence suggests that it isn't true, then, rather than restarting their life with a new set of beliefs, they will often cling more fervently to the old ones, generating new explanations for the conflict in their reality. Festinger called this response 'cognitive dissonance' and it occurs repeatedly in both the UFO field and many other systems of belief.

A genuine curiosity, *UFOs: Past Present and Future* was narrated by *Twilight Zone* creator Rod Serling and presented a solid and serious case for UFO reality, backed up by appearances from Robert Friend, William Coleman and J. Allen Hynek. A modest success on its release in 1974, the film was resurrected in 1979 in the wake of the colossal success of Spielberg's *Close Encounters of the Third Kind*. Entitled *UFOs: It Has Begun*, this extended version features new material suggesting that the cattle mutilations were being conducted by ETs and includes an interview with Gabe Valdez describing his research around Dulce.

Bob Emenegger's involvement with the UFO story did not end there. In the mid-1980s he was approached by two men high up in the Defense Audio Visual Agency (DAVA): Bob Scott, its director, and his aide Glenn Miller, a retired army general who had worked with General George Patton and been Ronald Reagan's first Hollywood agent. Scott had previously worked for the US Information Agency, whose job was to broadcast pro-American propaganda overseas, predominantly in the Eastern Bloc. Once again it was Coleman and Shartle who set the wheels in motion and brought together Scott, Miller and Emenegger, and once again they promised that the Air Force was ready to unveil its UFO secrets to the public. After some strange meetings, during one of which Scott expressed his conviction that the Earth was being visited by several species of ET, the project was dropped.

In 1988, the Holloman landing story resurfaced on national television in a tacky production called *UFO Cover-Up Live!*, aired on 18 October and produced by a former colleague of Emenegger's from Grey Advertising, Michael Seligman, who had previously produced Oscar awards ceremonies and, according to Emenegger, was too interested in making money to be part of the UFO conspiracy. On the show, Bob Emenegger told his story, while Shartle confirmed that he had seen the Holloman landing footage.

One of the most striking sequences of the programme is an interview with two Air Force insiders who are given the names

Falcon and Condor. Their faces hidden in shadow, the insiders tell remarkable stories: about a treaty with an alien race; of live extraterrestrials held at Area 51 who enjoy eating vegetables and strawberry ice cream and listening to Tibetan music. They also mention an exchange programme, though they don't call it Serpo. And who were these feathered fiends? None other than Rick Doty, aka Falcon (though not, apparently, Bill Moore's *original* Falcon), and Robert Collins, aka Condor, who would reunite in 2005 as the authors of *Exempt from Disclosure*.

So what was the purpose, if any, behind *UFOs: Past Present and Future*? Did the Air Force hope to draw in curious new recruits by presenting itself as sympathetic to the UFO subject? Did they hope to take their psychological programme promoting belief in ET visitation to a wider arena? To encourage popular interest in UFOs while simultaneously denying their own involvement with the subject seems contradictory, but then so does much of the Air Force's behaviour when it comes to UFOs. Does the film reveal an internal schism on the matter or – more likely, given the backgrounds of Coleman and Weinbrenner in AFOSI and the Foreign Technology Division – did it have a counter-intelligence and disinformation aim?

It certainly made an impact: the crux of the film, the Holloman landing, would go on to have a long and distinguished career in the UFO field. The climax of *Close Encounters of the Third Kind* is an elaborate, disco remix of the alleged event, which also formed the centrepiece of the Serpo story thirty years later. But before then it had a key role to play in one more classic piece of AFOSI dirty work, and once again we find Rick Doty at the centre of the cyclone.

THE ET FACTOR

By 1983, Paul Bennewitz's attentions were focused predominantly on the alien base at Dulce. Thanks to his *Project Beta* report, he and his sensational tales of alien infiltration were now well known in the UFO community. At the same time, Doty, Moore and Bob Pratt were discussing their SF novel, *The Aquarius Project*, and some of the ideas they were bouncing around were being included in the faked UFO documents that AFOSI were seeding to Bennewitz and other researchers.

One of these was Linda Moulton Howe, the Colorado-based writer and director of the 1980 cattle mutilation documentary *A Strange Harvest*, for which she had won a regional Emmy. Howe had got to know Paul Bennewitz and Gabe Valdez through her mutilation research and was convinced that there was a UFO connection to the phenomenon, so when HBO approached her to make a film about UFOs, she jumped at the chance. Its title, *UFOs: The ET Factor* says everything about what HBO hoped to get their hands on: this was *Close Encounters* and *ET*, but for real.

In April 1983, Howe was invited out to Kirtland by Doty. When she arrived at Albuquerque airport, UFOs were big news: the front page of the local newspaper ran a story about the release of Air Force documents recording UFO sightings over the Coyote Canyon at Kirtland in 1980.[7] The paper also reported that local scientist Paul Bennewitz had given a talk about UFOs to the Alburquerque chapter of the Mutual UFO Network.

Doty failed to meet Howe at the airport as arranged, and when he did show up he appeared anxious and irritable. As they drove into Kirtland, Howe asked Doty about the Holloman UFO landing. Doty replied that it had happened, but that the correct date was 25 April 1964, hours after policeman Lonnie Zamora had seen an egg-shaped craft land in a dry gully just outside Socorro, New Mexico.[8]

Doty led Howe to what he said was his boss's office, and they sat down on either side of a desk. Howe, who was used to the company of sturdy ranchers and slick TV professionals, found Doty, dressed in civilian clothing, to be a rather unimpressive character, but he knew how to get Linda's attention. Her film *A Strange Harvest*, he told her, had upset people in the government; she was on to something important. He and others at AFOSI wanted to help her to get the truth out through her documentary. As a taster for what they could offer, Doty opened a drawer and handed Linda some documents: 'My superiors have asked me to show you this,' he said. He insisted that she read them in a chair placed against a large mirror away from the windows. 'Eyes can see through windows', he told her. Howe believes that somebody was on the other side of the mirror, watching her reactions.

What Howe was looking at was AFOSI's fabricated Aquarius document, and it blew her mind. Headed 'Briefing Paper for the President of the United States' it laid out the now familiar scenario of UFO crashes, including Roswell, and of a surviving alien, EBE, being kept alive at Los Alamos until its death in 1952. EBE told us that its race had been on Earth for thousands of years and were still here. They had genetically seeded human life and guided our development through spiritual leaders, including one 'who was placed here to teach peace and love' two thousand years ago. Much of this material stemmed from discussions that Bill Moore and Bob Pratt had been having about their book, ideas already popular thanks to Erich von Däniken's *Chariots of the Gods?*. Moore had passed the ideas back to Doty, who was now feeding them to Howe.

The Aquarius document stated that Project Blue Book had existed only to draw public attention away from genuine, secret ET technologies, and that matters concerning the real UFOs were governed by a body called MJ-12. It also mentioned the UFO crash at Aztec, largely forgotten since its promotion in Frank Scully's *Behind the Flying Saucers*, but a hot topic again in the late 1970s, thanks to some well-placed tipoffs from anonymous Air Force sources. Bill Moore had been digging into the case and discussing it with Rick Doty and Bob Pratt, and via Doty it made its way into the Aquarius document, and back out into the UFO community, now with the government seal of approval. The process was dizzying in its circularity.

Doty offered to provide Howe with film footage that would include crashed UFOs, ET bodies, the Holloman landing film and, most incredible of all, footage of the living EBE. He also suggested that she might be able to meet EBE 3, an alien currently residing as a guest of the US government, perhaps even to film it. There was a catch to all this, however. Howe's film was to tell the UFO story only as far as 1964, the year of the Socorro incident. Why he chose this cutoff date is not clear; perhaps it was to steer Howe away from Dulce and the Bennewitz material.

Howe was stunned. She was going to be the one to finally present the truth about UFOs to the world. But why was the Air Force going to give her the material and not the *New York Times* or one of the major TV news programmes? Doty replied, in a moment of candour that seems not to have rung any alarm bells with Howe, that an individual was easier to manipulate and discredit than a large organization.

Howe told HBO to prepare themselves for the story of the millennium, but her producer at HBO insisted that the US Air Force provide them with a legally binding letter about the promised footage. Doty assured Howe that she would get the letter. Weeks passed, then months. Doty promised to arrange an interview with the Air Force colonel who had looked after EBE 1, but he needed to run background checks on her and her crew first. Nothing happened. HBO meanwhile, were starting to get anxious; where was the footage? By October 1983, Howe's initial contract with HBO had expired and they dropped the film. Howe was devastated.

Some years later Doty would deny ever having had this exchange with Howe, who responded by signing a sworn statement that the discussions had taken place. She also provided documentation of her dealings with both Doty and HBO. Why AFOSI decided to deceive Linda Howe is unclear. Was it always their plan to shut down the HBO production or was this just an unexpected bonus? We may not know exactly why, but we know they did it and, twenty-four years later, Doty admits that the exchange took place, much as Linda Howe described it. Its purpose was connected to AFOSI's disinformation programme against Bennewitz. As Doty put it: 'We gave Linda some good information, and some bad information. She chose the bad information.' And that's all he had to say on the matter.

You might expect that, having been given the runaround so completely that it could have damaged her career, Linda Howe would drop the UFO subject entirely. Instead, following Leon Festinger's cognitive dissonance model, she became even more certain that the ET presence on Earth was a reality, and has since dedicated much of her career to documenting the fact.

It's a pattern that recurs time and time again in the ufological arena, which is so rife with contradiction and paradox that believers must regularly develop complex new reasonings to maintain their faith in an extraterrestrial reality. This is something the architects of the disinformation game knew only too well and, in the wake of Howe's deception, they were about to produce their masterpiece.

THIRTEEN

BAD INFORMATION

'If you give a man the correct information for seven years, he may believe the incorrect information on the first day of the eighth year when it is necessary, from your point of view, that he should do so.

A Psychological Warfare Casebook *(1958)*

'We gave Linda some good information, and some bad information. She chose the bad information.' Rick's words haunted me. As the Laughlin Convention came to a close I returned to the dealers' room where people were now offering their wares at reduced prices: so many books, so many magazines, so much knowledge. I wondered, with some sadness, whether it could *all* be bad information, and if so, where had it come from?

THE BLACK GAME

As a young printer during the Second World War, Ellic Howe found himself much in demand working for Britain's Political Warfare Executive (PWE) under Denis Sefton Delmer, a former foreign correspondent for the *Daily Express*. Delmer specialized in the black arts – conjuring up schemes to mislead and demoralize Britain's enemies – and considered himself more of an artist than a soldier. The most infamous of Delmer's projects was Gustav

Siegfried Eins, a radio station that broadcast the anti-Nazi tirades of a mysterious German aristocrat known only as *Der Chef* (the chief) into Germany from June 1941 until December 1943. *Der Chef*'s sensational rants mixed genuine information gathered by British intelligence agents with slanderous gossip about the Nazi high command, including tales of political corruption and sexual debauchery. Wrote Delmer: 'We want to spread disruptive and disturbing news among the Germans which will induce them to distrust their government and disobey it, not so much from high-minded political motives as from ordinary human weakness.'[1]

Working under Delmer, Ellic Howe specialized in print, his work ranging from faking stamps and ration cards to more elaborate publications, such as a 104-page booklet, widely circulated to German troops all over Europe, known as 'the Malingerer'. In it a Dr med. Wohltat (Dr Do Good) recommended numerous ways in which soldiers might feign illness and injury in order to avoid military service. These ranged from distilling herbs to produce dramatic symptoms, to faking doctors' prescriptions and acting out conditions from sprains to amnesia. The booklets must have left an impression on the enemy, because German-made versions began spreading among Allied troops soon afterwards.

One of the most unusual of PWE's print projects was *Der Zenit*, a fake astrological journal that ran for six issues between 1942 and 1943. Elegantly designed, it fed into the already superstitious Nazi mindset – Hitler and Himmler both employed personal astrologers – and questioned, on astrological grounds, everything from the choice of Hitler's physician, to the timing of U-boat launches. *Der Zenit* used real astrologers to create subtly slewed data, and interpretations of data, that were convincing enough to fool even the most ardent astrologer, at least for a while. The results of such black propaganda operations aren't always obvious, and may take months, even years to have a noticeable effect; however, once successfully sewn, a single seed of uncertainty can grow to become a forest of doubt.

Delmer's PWE was largely working from a blank slate, but in the years after the Second World War disinformation techniques became increasingly sophisticated. During the Cold War both the KGB and the numerous American intelligence agencies employed their own specialized forgery divisions. It's estimated that the Soviets were spending about $3 billion a year on disinformation activities, while the US spent considerably more, with the CIA, just one agency, dedicating $3.5 billion to the black and grey arts.

A favourite tool employed by both sides was planting news stories, either by having your agents in place within news organizations at home and abroad or, in less familiar territories, by bribing journalists and editors. With the growth of the networked global news media, this meant that a story planted in Nigeria might seep its poison a few months, or even years, later, into the heart of your adversary's media.

One of the most successful applications of *aktivnyye meropiyatiya*, or 'active measures' by 'Directorate A', the KGB's disinformation specialists, started life as a letter in *Patriot*, India's pro-Soviet daily newspaper, on 16 July 1983. Titled 'AIDS May Invade India: Mystery Disease Caused by US Experiments', the letter was written by an anonymous 'well-known scientist and anthropologist'. It suggested that the AIDS virus had been developed by Pentagon biological warfare specialists at Fort Detrick, Maryland, as an 'ethnic weapon' targeting blacks and Asians.

It took two years for the story to gain a foothold, after an October 1985 article in the Soviet weekly *Literaturnaya Gazeta* quoted the letter from the 'well-respected' Indian newspaper, *Patriot*. This new variant on the story added that the US was breeding AIDS-carrying mosquitos and infecting its servicemen in foreign countries. But the masterstroke came in 1986, at an AIDS conference in Zimbabwe, where three East German scientists, Professor Jacob Segal, former Director of the Institute of Biology in East Berlin, his wife Dr Lili Segal and Dr Ronald Dehmlow, presented a paper on the subject, lending it the academic authority it needed to be considered seriously.

It worked. By August 1986, the idea was being discussed openly in the *Journal of the Royal Society of Medicine* and, on 26 October, it hit the front page of Britain's *Sunday Express*: 'AIDS Made in Lab Shock'. From this point, the story grew out of control. Photographs of an ethnic weapon being fired appeared in a Kuwait newspaper, *Al-Oabas*, and were picked up by Reuters, hitting the big time on America's CBS Evening News in March 1987. What had begun as an obscure item in an Indian newspaper was now being broadcast into millions of American homes – and millions of American minds.

The Soviets always denied creating the story, though in late 1988, under pressure from the Americans, they agreed to stop actively promoting it. But by then the damage was done; the ethnic weapon now even had a codename, 'Plus Morbific Plus'. While it would largely disappear from the mainsteam media, the AIDS weapon story had become a staple of the underground conspiracy culture, and still thrives on the Internet.

Despite their earlier denials, following the collapse of the Soviet Union, the KGB came clean. On 19 March 1992, in the Russian newspaper *Izvestiya*, Russia's intelligence chief Yevgeni Primakov (who became Prime Minister in 1998) admitted that the original story was a Directorate A special: 'the articles exposing US scientists' crafty plots were fabricated in KGB offices'. We can assume that Plus Morbific Plus was just one of many such tales cooked up by disinformation artists on both sides of the Cold War.

And what, you might ask, does this have to do with UFOs? The answer is that it has *everything* to do with UFOs. In the mid-1980s a combination of PWE's document counterfeiting techniques and the kind of targeted propaganda employed so effectively by KGB Directorate A would result in a near-fatal assault on the cohesion and credibility of America's now-thriving UFO community, perhaps the most devastating salvo in the Air Force's ongoing information war.

MJ-12

On 11 December 1984, a manila envelope with an Albuquerque postmark dropped through the door of Jamie Shandera, a TV producer working with Bill Moore on his UFO investigations.

Inside was a roll of 35mm film containing photographs of a paper, dated 18 November 1952, prepared by the first CIA director Rear Admiral Roscoe Hillenkoetter, briefing incumbent President Dwight Eisenhower on the existence of Operation Majestic 12. The Majestic or MJ-12 group consisted of twelve specially selected scientists, military and intelligence specialists brought together following the 1947 Roswell UFO crash to study the wreckage and its occupants. Projects Sign and Grudge, the document stated, had been set up to collect data to aid the group's work. The document referred to the upsurge in UFO sightings that took place in 1952 and concluded that the MJ-12 project should continue into the new president's administration with the 'imposition of the strictest security precautions'. A list of attachments was included with the briefing document, along with a note signed by Harry Truman, dated 24 September 1947, instigating the project.

Moore was, by now, well used to getting the runaround from government UFO documents. The Majestic 12 group were also familiar to him from the Aquarius document that Rick Doty had asked him to give to Paul Bennewitz, the same document that would have formed the basis for their book. Were these new documents the stash that the previous few years' trail of bread-crumbs had been leading him to? Was this the motherload?

Moore decided that the best thing to do with the documents was to sit on them, though he did give copies to a select few fellow ufologists, including fellow Roswell researcher Stanton Friedman and Lee Graham, an aerospace defence contractor. Perhaps this delay frustrated the source of the original package because, over the following months, Moore and Shandera received a series of postcards, this time postmarked New Zealand, with a return address of Box 189, Addis Ababa, Ethiopia. The cards bore short phrases like 'Reeses Pieces' and 'Suitland' that meant nothing to either of the recipients. Then, in what seems like a remarkable moment of serendipity, Friedman, a Canadian resident, asked Moore and Shandera to visit a collection of recently declassified US Air Force documents at the National Archives. The archives were based at Suitland, Maryland. The head archivist there was called Ed Reese.

Sure enough, in Box 189 of the new document collection, Shandera and Moore found what appeared to be confirmatory supporting evidence for the MJ-12 papers. The smoking gun was a memo, dated 14 July 1954, in which Robert Cutler, Executive Secretary of the National Security Council, advised Nathan Twining, the Air Force Chief of Staff and alleged MJ-12 member, of a rescheduled MJ-12 meeting.

Although it lacked an archival catalogue number, the memo was printed on the sort of onionskin paper that would have been used for carbon copies at the time, lending it greater authenticity than the roll of film on which the first batch of documents had appeared. Unusually for a carbon copy of the period the paper had been folded, as if enclosed in a shirt pocket.

Someone had clearly wanted Shandera and Moore to find this document, with the hope of bolstering their enthusiasm for the initial MJ-12 release. But who? And how had they smuggled it in? Security at the National Archives is tight, making slipping a document into the system difficult, though not impossible. If it was a forgery, perhaps it had been introduced to Box 189 before being deposited with the archives, but if this was the case, why did it not have a catalogue number like everything else?

Moore, Shandera and Friedman kept MJ-12 close to their chests, though word of its existence began leaking out to the inner circles of the UFO community. And somebody wanted it to go further. In 1986, Britain's best-known UFO researcher, Jenny Randles, was approached by an anonymous source offering her evidence of an American government conspiracy to cover up evidence of UFOs. Wary of being taken for a ride, she declined the material.

In early summer 1987, another up-and-coming British UFO researcher, Timothy Good, was less picky and took the bait, presenting the MJ-12 papers as an appendix to his book *Above Top Secret*, due to be published that July. Word filtered back to Bill Moore, who decided that it was time to go public, which he did at the National UFO Conference on 13 June. The mainstream press immediately picked up the story, which within days had featured on ABC TV's high-profile *Nightline* news programme, and in the *New York Times*.

Like the KGB's AIDS hoax, MJ-12 had taken three years to travel from remote obscurity to the national news. Timothy Good's *Above Top Secret*, meanwhile, became an international bestseller and put UFOs once more on to the public agenda. This was disinformation as it is supposed to happen. The MJ-12 papers caused huge fractures across the entire UFO community, pitching those who believed the documents to be the evidence they had all been waiting for against those who felt that they were just another hoax. Twenty-five years later the battle lines are still intact, while reams of further alleged MJ-12 documents continue to surface, creating new dissensus and confusion.

In 1988, the FBI's counter-intelligence division were asked to investigate the MJ-12 papers on behalf of AFOSI. One of their trainees was found in possession of a copy: with their 'Top Secret' markings, this might have represented a disclosure of classified information, constituting a serious federal offence.

The FBI's protracted investigation is related by investigative journalist Howard Blum in his book *Out There*. FBI agents showed copies of the papers to numerous government agencies, none of which knew anything about them. They then considered their enemies, the Soviets and the Chinese. The FBI learned that the CIA had spread UFO stories into hostile territories as part of its Cold War operations,[2] so perhaps the MJ-12 papers were a form of retaliation? They could find no evidence to support this particular theory.

The FBI then set its sights on AFOSI itself, specifically the team at Kirtland, but once again they had no luck. As Blum relates:

> Every one of them, to a man, firmly denied he had a role in writing the MJ-12 documents. And to complicate matters further, many of the agents had suddenly decided to retire. They were now private citizens with, they emphatically emphasized (and on at least one occasion a lawyer was brought in to reiterate), all the constitutional rights of private citizens.[3]

An exasperated FBI agent concluded: 'We've gone knocking on every door in Washington with those MJ-12 papers. All we're finding out is that the government doesn't know what it knows. There are too many secret levels . . . It wouldn't surprise me if we never know if the papers are genuine or not.'[4]

In the end, the answer came from AFOSI: the document, they admitted, was 'completely bogus'. If AFOSI did know who was responsible for the MJ-12 papers, they weren't telling. We do know that Rick Doty was among those questioned as part of the investigation, and we also know that he left the Air Force in 1988, becoming a private citizen, having already been asked to leave AFOSI following an unspecified incident at Wiesbaden AFB, West Germany, in 1986.

The papers fit neatly into Kirtland AFOSI's disinformation campaign against the UFO community. The first known mention of MJ-12 came in November 1980, at the end of the Aquarius document seen by Bill Moore at Kirtland. The only version of this document in the public domain is the one that Moore had himself 'retouched' before feeding it to Paul Bennewitz, Linda Howe and others. Its final paragraph reads: 'The official US Government Policy and results of project Aquarius is [sic] still classified top secret with no dissemination outside official intelligence channels and with restricted access to "MJ Twelve".'

In notes from a January 1982 conversation with Moore, *National Enquirer* journalist Bob Pratt sketched out an early version of the MJ-12 idea that would eventually be included in their unpublished novel. 'Govt . . . UFO Project is Aquarius, classified Top Secret with access restricted to MJ 12 (MJ may be "Majic")'. Moore and Roswell researcher Stanton Friedman, meanwhile, had speculated at length as to who might make up the Majic team of twelve outlined in the original Aquarius memo: surprise, surprise, eleven of them appeared in the papers later mailed to Jamie Shandera.

Beyond these issues of provenance there are multiple technical and factual problems with the documents, the most heinous being the location of the alleged Roswell UFO crash site – MJ-12's *raison d'être*. The original briefing document states that the ranch where the ET craft was found is 'approximately seventy-five miles north-west of Roswell Army Air Base'. In fact it is sixty-two miles by air and more than a hundred miles by road. Nit-picking you might think, but surely the keepers of the world's greatest secret, when reporting to their new boss, would make sure that they had their facts straight. Funnily enough, in William Moore and Charles Berlitz's *The Roswell Incident*, the first book to be written about the crash, the same distance is listed as . . . seventy-five miles.[5]

So who created the MJ-12 documents? The finger of suspicion points unwaveringly at AFOSI, using information inadvertently provided by Bill Moore, Bob Pratt and Stanton Friedman. But even then there are questions: if AFOSI were behind the fakes, could they really have been so negligent as to post them from Kirtland AFOSI's home city of Albuquerque?

We can only assume that whoever posted them – and these aren't necessarily the same people who created them, or photographed them – wanted the package to look as if it came from the base. Perhaps, as Moore has suggested, he was supposed to believe that this was his reward for all his hard work on AFOSI's behalf; or perhaps whoever created them didn't actually care that they would be exposed as bogus – maybe that wasn't the point. It's clear that the MJ-12 documents were targeted at the global UFO community – when Moore and Shandera didn't go public with them, the files were offered to two British ufologists – and that they were intended to keep people believing in UFOs. But why?

SECRECY AND STEALTH

The timing of the MJ-12 release, the climax of the operation that began with the deception of Paul Bennewitz, ties in closely with the early flights of the F-117A 'Nighthawk' Stealth Fighter, then one of the US Air Force's most secret technological assets. The decision to go into production of the Stealth was made in 1978, the year of the Ellsworth hoax and the year that tales of UFO crashes began to be leaked from the Air Force to ufologists, leading to the resurrection of the Roswell story. The first flying model took to the air in June 1981, while the first fully operational craft flew out of Area 51 in October 1983, remaining super-secret until 1988. These first Stealths were tested in Nevada, at Area 51 and the neighbouring Tonopah Test Range, so what better way to keep aircraft spotters and UFO hunters off its trail than to lure them out to Kirtland or Dulce, and then distract them with paper chaff in the form of the MJ-12 documents?

A tangled paper trail would seem to back up these assertions. One of the people Bill Moore was instructed to investigate during his allegiance with AFOSI was Lee Graham, who worked building parts for the Stealth Fighter and defence satellites at Aerojet Electrosystems in Azusa, California. Graham and his Aerojet colleague Ron Regehr also shared a fascination with UFOs, a topic that was always hovering at the edges of their work in the classified technology field. Curious to know what the aircraft that they were helping to build looked like, Graham and Regehr used to go out 'Stealth hunting' at Tonopah, hoping to catch a glimpse of the Nighthawk in flight – something that they were well aware, as contactors, they were not supposed to be doing.[6]

Lee Graham first contacted Moore after reading *The Roswell Incident* and they exchanged data about UFOs. This must have raised eyebrows at AFOSI because before long Moore was feeding Graham the Aquarius document and, later, the MJ-12 papers. One day Graham asked Moore how he had obtained these documents, so Moore showed him a Defense Investigation Service (DIS) badge – DIS (now the Defense Security Service) dealt with industrial security on government projects. Graham became suspicious of Moore, worrying that if this went any further then his own secret clearance might be at risk and he might lose his job; so Graham showed the UFO documents to Aerojet's head of security and suggested that they look into Moore's activities.

Instead it was Graham who found himself under investigation, eventually being paid a visit by an FBI agent and a man in civilian clothing. The men strongly dissuaded Graham from continuing his Stealth-chasing activities, but actively encouraged him to keep talking about UFOs and disseminating the MJ-12 papers – contradictory instructions reminiscent of those given to Olavo Fontes and Silas Newton back in the 1950s.

Graham later discovered that the man with the FBI agent was Major General Michael Kerby, Director of the Air Force Legislative Liaison office, responsible for Stealth Fighter flights and other secret projects out at Area 51. Years later, Graham got hold of his own DIS investigation report, which noted that, while he was no traitor, he could be tricked into handing over classified data, especially to fellow UFO enthusiasts – such as Bill Moore.

One other important name comes up in Graham's security files: Colonel Barry Hennessey. Hennessey's Air Force biography states that he was a member of the Defense Intelligence Senior Executive Service and Director of Security for Counter-intelligence and Special Program Oversight at the Office of the Secretary of the Air Force. As such he was 'responsible for security and CI [counter-intelligence] policy as well as management oversight of all Air Force Security and Special Access Programs, which includes ensuring the security of various research projects with significant potential impact on the defense capabilities of this country'.

Previous to his role as Director of Security, during the period of the MJ-12 release and the Bennewitz affair, Hennessey ran AFOSI's PJ or Special Projects unit at the Pentagon. A 1989 letter sent to Lee Graham describes PJ's role as follows:

> [PJ] Develops and implements security policies and procedures for compartment Special Access Programs such as the B-2 and F117-A aircraft. [PJ] Guides and directs security-related activities of all program participants, both in government and industry. [PJ] Provides counter-intelligence and security support to special Air Force activities. [PJ] Operates two detachments in classified locations.

Was Hennessey's PJ department responsible for the operation that brought together Rick Doty and Paul Bennewitz? Did they concoct the Aquarius documents and the MJ-12 papers, using information provided by Doty and gleaned from Moore? At the moment, it's the best answer that we have. So does this mean that Hennessey was the Falcon, the man in the red tie who first contacted Bill Moore and introduced him to Doty? While it would make perfect sense, both he and Moore have denied it, leading researchers to look elsewhere, including the unlikely possibility that there was no Falcon beyond Doty himself. Doty, remember, appeared as 'Falcon' in the *UFO Cover-Up Live!* TV programme of October 1988.

Whoever the Falcon may have been, the chain of command here points to an operation authorized from the top down, with Rick Doty acting as an agent 'in the field' somewhere near the bottom. The entire Doty–Moore–Bennewitz–Graham episode reveals a great deal about the methods and concerns of the US Air Force in protecting their secret assets, and also in their attitudes towards the UFO community. To the Air Force, ufologists are a nuisance, and sometimes a necessary nuisance; but when they work with sensitive data, like Graham and perhaps Paul Bennewitz, then they need to be closely monitored, tested, occasionally exploited and sometimes neutralized.

For many researchers the MJ-12 papers are the holy grail of the UFO conspiracy, demonstrating once and for all that not only did the Roswell UFO crash happen, but that the cover-up of the truth was instigated in its wake. At the heart of the matter is a peculiar paradox: the UFO community repeatedly states that the US government has lied about the truth, and yet government documents, when presented in the right way, become hard evidence and a legitimization of that truth.

For the faithful, the MJ-12 documents demonstrate that somewhere in the labyrinths of power, somebody knows that the extraterrestrial UFO phenomenon is reality; this means that one day there will be disclosure, a revelation. This disclosure has been a beacon of hope for the UFO community since the days of Donald Keyhoe; it cannot, and most likely will not, ever be extinguished. Keeping the beacon aflame will be as important to whoever is doing Hennessey's job today as it was in the 1980s, or the 1950s; because if there are no more UFOs, then there is no more cover for the Pentagon's many Special Projects.

That the data trail so clearly links AFOSI to the MJ-12 documents appears not to have dissuaded the UFO party faithful, who continue to promote MJ-12 and the Roswell crash to this day. Here, again, we see a collective cognitive dissonance in action: a profound denial of the reality of the situation in the face of increasingly obvious facts. We can also see a form of the Stockholm Syndrome at work here, in which hostages become emotionally attached to their captors. The MJ-12 papers have held the UFO community captive since 1984 and represent their only hope of proving a government UFO conspiracy; even some of those who accept that the documents are bogus still defend them as disinformation pointing to a deeper truth. And the documents keep on coming. Since the early 1990s, like a slowly expanding alternate universe, or expansion packs for a role-playing game, new batches of papers have appeared, generating further layers of complexity to the MJ-12 scenario. Facsimiles of these new documents can be found online, the most imaginative being the 1954 *Special Operations Manual (SOM1-01)*, a briefing guide for those involved in the recovery of extraterrestrial spacecraft.

If the purpose of the MJ-12 documents was to distract, fragment and weaken the UFO community, then they succeeded beyond the wildest dreams of their creators, and should be considered a masterpiece of the black arts. Whoever was responsible has launched a mythos to rival anything imagined by J. R. R. Tolkien or George Lucas, and one that is sure to endure for just as long.

ALIEN SEED

In the second half of the 1980s the information contained in the MJ-12 papers, the documents circulated by Bill Moore and Rick Doty, and in Paul Bennewitz's *Project Beta*, began to escape the initial closed circle of ufological insiders and infect the wider culture. The Internet was now twitching into life, with text-only dial-in bulletin boards serving as perfect repositories and transmission points for these wild tales.

While some serious American UFO researchers immediately recognized early on the potential of the Internet as a resource-gathering and sharing tool, it also drew the attentions of the newly emerging culture of wired-up computer geeks and hackers, many of whom – and I count myself among them – had grown up viewing the world through lenses coloured by role-playing and video games, and fantasy, horror and science-fiction literature. This new breed of enthusiast, with their feet planted in the bundles of cables beneath their computer terminals and their eyes turned upwards to the stars, were perfectly primed to receive the richly complex, superficially detailed and sensationally bizarre patchwork of information streaming out of the UFO culture.

Probably the most influential source of information in this period was John Lear. His estranged father was William Lear, inventor of the Lear Jet and the eight-track tape recorder, who had spoken publicly about his belief that flight within and beyond the Earth's atmosphere could be achieved using electromagnetically controlled gravitational fields. For many years John Lear worked as a professional pilot, flying Air America missions for the CIA in south-east Asia, among other jobs. Regarded as something of a hotshot flyer, his name and reputation as an Air Force and intelligence insider made him an impressive source when he began distributing UFO information in late 1987.

Lear first got the UFO bug earlier that year and from there began hoovering up as much information as he could, reading books, watching films and speaking to Paul Bennewitz, Bill Moore and others. Lear's first public statement on the UFO situation, widely transmitted over various bulletin boards, is a perfect synthesis of the Aquarius and MJ-12 disinformation and the chthonic, paranoiac horrors of Paul Bennewitz.

An online interview with Lear, from 14 February 1988, gives a sense of the kinds of material he was disseminating – much of which will now be familiar to us: the Strategic Defence Initiative was being developed not to fight off Russian missiles, but to protect us from ET attack; Ronald Reagan and Mikhail Gorbachev's meetings in Iceland were held to discuss the ET threat, a major factor in the Cold War thaw; CNN were about to broadcast a bean-spilling interview with a member of MJ-12 and footage of an ET being telepathically interviewed at Los Alamos by an Air Force colonel; another video showed an ET device projecting images filmed at Christ's crucifixion; meanwhile, from their base beneath Dulce, the ETs were abducting humans and animals in order to conduct horrific genetic breeding experiments, captured ET craft were being flown by the government at Area 51 . . . and so it went on.

By the time of Bill Moore's confession at the 1989 MUFON conference, the material promoted by Lear was bringing new blood into the UFO community, while simultaneously tearing it apart. Sensational, terrifying and almost entirely absurd, it was the stuff of science-fiction nightmares, which is exactly what it became. The stories dreamed up by Bill Moore and Rick Doty and delivered to Paul Bennewitz fanned their way out through these new conduits and consumed new minds – some of them, no doubt, deep inside the government and the Pentagon.

Every once in a while the material would surface into the mainstream media through newspapers, magazines, TV news or non-fiction programming like *UFO Cover-Up Live!* or *Unsolved Mysteries*. In the mid-1990s, ten years after leaving AFOSI's stable, these ideas transmuted once again, this time into fiction. *The X Files*, *Dark Skies* and the blockbuster film *Independence Day* were extremely effective vehicles for their transmission, having as much of an impact on the popular imagination as had Spielberg's *Close Encounters* two decades previously.

Another decade on, at the Laughlin UFO Convention, the same stories were still being told. There are new MJ-12 documents, new videos, new witnesses and new whistleblowers, and they are all still drumming out the same message: the ETs are real, they are here, and they are talking to the American government. It's a message that has been repeated and retold so often over the years that it has begun to carve permanent tracks into the memories of those who hear it, shedding their uncertainties and taking on the familiarity of truth.

Now, however, the convention was over. Most of the attendees would be returning to their home planets, at least until the next such meeting. John and I were relieved to be leaving the madness of the past week behind us, though we were sorry to say goodbye to Bill Ryan. Despite the week's stresses, Bill was flush with confidence and a renewed sense of purpose following his first stint in the public eye. And his new job was not without its perks. Rather than returning to the UK as planned, Bill had been invited to spend a few days in Las Vegas with the attractive dark-haired woman who had approached him after his talk. 'You won't believe this,' he told us excitedly, 'but she's only got one human parent; the other is extraterrestrial.'

We wished Bill luck and advised him to keep his wits about him. None of us knew it yet, but his time as ambassador to Serpo was nearly over; an even grander mission awaited.

John and I, meanwhile, were headed to Albuquerque, where Rick had agreed to meet us for an interview.

FOURTEEN

FLYING SAUCERS ARE REAL!

'If we persist in refusing to recognize the existence of the UFOs, we will end up, one fine day, mistaking them for the guided missiles of an enemy – and the worst will be upon us'

General L.M. Chassin, NATO co-ordinator of Allied Air Service, 1960

Albuquerque Airport is bolted on to Kirtland Air Force Base, which covers territory about half the size of the city itself. Some of the base's more distinctive features are visible as you come in to land, including a large wooden ramp on a horizontal U-bend that looks like an old-fashioned rollercoaster, but is used for testing the effects of electromagnetic fields on aircraft.

Like the base, Albuquerque itself is large and sprawling, with no distinct centre. Old Route 66, now Interstate 40, cuts across the city east–west, passing through the mighty peaks of the Sandia mountains that halt the spread of housing at the east end of town. The Sandias dominate the skyline, especially in the evenings, when they glow a deep, radioactive orange, appropriate for a city that houses the National Museum of Nuclear Science and History and one of the largest nuclear arsenals in the world.

We'd booked into a Travel Lodge at the south end of town, situated among pawn and gun shops, the boarded-up Atomic Motel and other remnants of a bygone era. We liked the motel, but Rick was dismayed to find us there.

'You shouldn't be staying here,' he told us as he pulled up in the parking lot. 'We call this area the war zone!'

Aha, that would explain the crack pipe our crew found down the back of one of their beds.

'So, do you guys want to see the base?' Rick asked.

'Of course we do!'

'Then get in the car!'

As we pulled up at Kirtland's entrance we passed the sign for Thunder Scientific, now run by Paul Bennewitz's sons; if anything stirred in Rick he didn't show it. At the guardpost I wondered if John and I would pass muster and be allowed inside but, with both of us wearing army-green jackets and, for me, unusually short hair, we looked more militarized than Rick, whose knitted sweater wouldn't have looked out of place on a Jim Reeves album.

An armed guard stepped up to the driver's side window. Confronted by gun-toting Americans in uniform, I always get the sense that I must be doing something wrong, but the soldier ignored John and me. Rick passed him an ID card and the guard saluted, the gate was raised and we drove in. I asked if I could see Rick's card, but he slipped it back into his wallet without saying anything.

Our first stop was Rick's old AFOSI office. It didn't look like much. Drab, low and long, a glorified Portakabin that could just as easily have housed a pest-control firm. Rick was openly proud of his time with AFOSI, and complained that the unit at Kirtland, now located in another building, had been cut back from more than fifty personnel to less than twenty since the end of the Cold War.

Driving around we got an unclassified glimpse of life on the base, a small industrial town in its own right. We passed a man in a t-shirt taking notes on a clipboard beside an enormous parabolic dish, its central antenna aimed into a large hangar the door of which was opened just enough to let in whatever it was transmitting, but not enough to let us see what was inside. The rows of

Quonset huts, hangars and office buildings revealed little of the work that went on inside them, except maybe the impressive-sounding Directed Energy Directorate (DED), which had its sci-fi name spelled out in sci-fi lettering on its front. Perhaps the world's leading centre for research into the military use of lasers, microwaves and plasmas, the DED's work includes developing lasers designed to be attached to aircraft and shoot down missiles. Another DED project is the Starfire Optical Range, a $27 million high-powered telescope that specializes in tracking satellites. Its 'adaptive' optics allow it to stay focused on a basketball-sized object a thousand miles away in space, regardless of atmospheric conditions. It maintains its focus by firing lasers into the night sky, lasers that the DED is now hoping to convert into satellite-disabling weapons.[1]

We drove away from the inhabited part of the base and out towards the Manzano foothills, where Paul Bennewitz had filmed UFOs from his home less than a mile away. Rick told us that the SAS and other Special Forces groups used the open desert and mountain terrain at Kirtland in training exercises, and looking around one could easily imagine oneself in Afghanistan or Pakistan. Paved roads headed up and around the isolated upswelling of rock at a shallow gradient, leading to entrances concealed around the back of the mountain in Coyote Canyon, the site of alleged UFO incidents in 1980. The Manzano mountainside was littered with reflective material, some of it radar-reflecting chaff, some of it possibly the remains of air accidents, like the B-29 that crashed there in 1950, killing all thirteen occupants – luckily without detonating its atomic cargo.

A formidable triple-electrified fence surrounded the mountain. Rick told us that it had claimed more than one life over the years, though it is no longer live. The security was warranted. In 1947 construction began on the Manzano Weapons Storage Area, a nuclear-blast-proof shelter cut deep into the mountain rock. Until the appearance of the first Soviet thermonuclear device in 1953, the mountain also contained a presidential bunker designed to protect Dwight Eisenhower from an atomic attack. A total of four separate underground plants were cut into the mountain, containing 122 magazines for weapons storage and other areas housing research facilities. Until 1992 the weapons magazines held a significant portion of the US's nuclear stockpile, though the warheads are now stored in a specialized underground facility elsewhere on the base. Today Manzano's tunnels are used as research facilities by Phillips Laboratory, one of Kirtland's main contractors, and as shooting locations for Hollywood movies.

It's hard not to feel excited when confronted with a *bona fide* secret – or at least once-secret – underground base. The frisson of so many childhood doodles and *Thunderbirds* fantasies rushed up my spine like a small underground nuclear test, making the hairs on the back of my neck rattle with the force of its detonation. For a moment I imagined another life in which I was the scientist or secret agent that at one time we all wanted to be, tapping an access code into a keypad alongside one of the discreet concrete entrances, large enough to get a small vehicle into, that we could see on the north-western flank of the mountain base. Inside I walk down another corridor, pass another palmprint-activated blast door and nod a greeting at a white-coated boffin tinkering with the wires of a nuclear weapon. He gives me a thumbs up in response. Another steel door swings open only after a retina scan and another access code has been entered. I pass through. It closes behind me. Another blast door opens and there it is, my baby, suspended on steel cables from the ceiling: a gleaming flying saucer with a mighty dent in its side from where it fell to Earth in 1947 . . .

'Well, I guess that's about all I can show you without getting you arrested,' said Rick, turning the car back to the main buildings of the base. 'Let's get you back to your hotel.'

We drove back past the rows of unmarked hangars and anonymous low-slung buildings. For me this was a place of wonder and mystery. To the people who spend every day here, it's a workplace, and perhaps some of them know the secrets that Paul Bennewitz was punished to protect.

As we reached the main entrance and the nondescript Thunder Scientific building, a battered Ford Transit van with tinted windows slid in front of us. I wouldn't have been surprised to see a long-haired heavy-rock band burst out from its bumper-sticker-adorned back doors, guitars shredding. Prominent among the stickers was the iconic black and white outline of a grey alien face.

I realized that Paul Bennewitz was right: the aliens were here at Kirtland, and had been all along.

DRAGON LADIES AND LIGHTNING BUGS

On 1 August 1955, at the then super-secret Groom Lake test site in Nevada (now better known as Area 51), the first prototype U-2 aircraft leapt suddenly into the air. The flight of the Dragon Lady, as the plane was known, opened a new era in aerial reconnaissance and a whole new series of headaches for the Air Force and, specifically, its public UFO investigation, Project Blue Book.

Until Blue Book's closure in 1969, a significant number of the UFO reports that it received – CIA historian Gerald Haines suggests as many as half[2] – were actually sightings of the U-2 and its stealthy successors, the CIA's A-12 Oxcart and the Air Force's SR-71 Blackbird, all operated by the CIA's Office of Scientific Intelligence. In the early days the silver U-2 would reflect sunlight, particularly at sunrise and sunset, making it quite visible from below, so triggering a few head-scratching phone calls to Blue Book. All three aircraft flew at 85–95,000 feet, almost three times the altitude of commercial airliners, while the Blackbird reached incredible speeds of Mach 3.2 (2,200 m.p.h.), once getting from London to New York in two hours. The Oxcart and Blackbird were also fitted with Palladium transmitters which might, on occasion, have been used to confuse the radars of nearby civilian aircraft.

It's no wonder that pilots and people on the ground imagined they were seeing something out-of-this-world. Presented with sightings of these 'UFOs' Blue Book's investigators would compare their reports with known reconnaissance overflights, then had the thankless task of convincing the witnesses that what they had seen was the planet Venus, a weather balloon, a temperature inversion, or whatever seemed convincing at the time.[3]

The venerable U-2 is still operational while the Blackbird went out of service in 1999. So what's doing the CIA and the Pentagon's advanced reconnaissance work today? Rumours persist of next-generation craft such as the fabled SR-75 'Aurora' (like the Stealth, the subject of a Testors model kit) and the triangular TR-3B, which may or may not be at the root of the many 'flying triangle' UFOs seen regularly all over the world. On 30 March 1990, for example, multiple witnesses saw and photographed a triangular craft with lights at each corner over Brussels, Belgium; the Belgian Air Force sent four F-16 jet fighters in hot pursuit, but they were outrun and then baffled by what sound like Palladium-style radar evasions. A decade later, on 5 January 2000, a large, slow-moving triangular craft was watched for nearly an hour by five policemen in the region of Scott AFB in Illinois. Few doubt that the flying triangles are real aircraft; perhaps they'll be formally introduced to the general public soon.

These days, however, most reconnaissance work is carried out more efficiently and less riskily by orbital satellites and unmanned aerial vehicles (UAVs), or drones. UAVs are presented as the future of warfare, their roles constantly being extended as their technology improves, but they aren't exactly a new idea. During the First World War a British inventor, Archibald Low, who had also demonstrated the first television broadcast, worked on an explosive radio-controlled aircraft known as the Aerial Target: a device that could have changed the way future battles were fought, but development was halted on this, and a wireless-guided missile, once the war was over. In the 1930s, the British and American Navies began converting old aircraft into 'drones', the name originating with a de Havilland Tiger Moth that became the Queen Bee.

Initially these pilotless planes found themselves in the rather inglorious role of targets for anti-aircraft gunners, but they soon proved useful in all sorts of situations.

In the early 1950s, Project Badboy saw the US Air Force converting F-80 jetfighters into remote-control drones, designated QF-80, a number of which were flown through mushroom clouds to collect radioactive air samples. During the 1953 Upshot-Knothole tests at the Nevada test site, several unfortunate monkeys were exposed to high radiation levels inside QF-80s. It's not inconceivable that crashes of monkey-carrying aircraft in such tests might have led to tales of UFO crashes and dead aliens.[4] Maybe the rumours spread via Scully's *Behind the Flying Saucers* were intended as a pre-emptive cover for Air Force experiments like these, in case they should end up flying off-range. Might such an incident lie even behind the Roswell story?[5]

Following the disastrous shootdowns by the Soviets of Francis Gary Powers's U-2 and a Boeing RB-47 reconnaissance aircraft, the military and intelligence services recognized the urgent need for crewless aircraft that would save pilots' lives and spare their generals from further embarrassment. The US Air Force asked the Ryan Aeronautical Company to develop purpose-built drones for them, the most successful of which was the Ryan 147. The 147s, known as Lightning Bugs, came to the fore as reconnaissance craft during the Cuban Missile Crisis, saw regular action in Vietnam and were still being used in the 1990s. Even the earliest models boasted impressive specifications: programmable flight paths, a cruising altitude of 55,000 feet, a top speed of 720 m.p.h. (about Mach 1) and a range of 1,200 miles. Later models far exceeded this: introduced in 1966, the Lockheed D-21 Tagboard, a CIA relative of the Lightning Bugs, hit a staggering 2,700 m.p.h. (Mach 3) at an altitude of 95,000 feet, but was dangerously unstable and flew only four official missions. Despite wingspans the size of a small plane, the Lightning Bugs were able to evade radar detection and, once modified not to leave a contrail, were almost invisible to the naked eye.

The 147s could be kitted out to perform all sorts of nifty tricks. The 147D, for instance, doubled up as both a target and a reconnaissance craft. It would fly over enemy territory, drawing out surface-to-air-missiles (SAMs), then receive and transmit signals from both the launcher on the ground, and the missiles themselves, in the moments before being blown to smithereens. This vital signals intelligence helped to map the locations of SAM sites and develop effective counter-measures against new guided-missile technologies. It was a suicide mission that no human pilot would want to take on, but it was all in a day's work for the selfless Lightning Bug.

In a variation of the Palladium technology, drones were fitted with 'travelling wave tubes', effectively radar amplifiers, that generated returns much larger than they actually were, making missiles more likely to miss their targets; they may also have frightened pilots and radar operators into thinking that a huge aircraft was in the vicinity – as described by Milton Torres over Kent in 1957. The device could also be tuned to mimic the radar returns of specific aircraft. Another device, the 'shoe horn', turned the Lightning Bugs into radar ventriloquists, able to throw their radar signatures to locations other than their own, suggesting how UFOs might appear to 'jump' across radar screens at impossible speeds – as reported during 1990's flying triangle chase over Belgium. Improved gyroscopes in later models allowed them to perform sudden turns when targeted by enemy radars, again fitting the classically erratic flight path of so many UFOs. Drones could be programmed to fly at altitudes as low as 150 feet – one allegedly photographed an electrical pylon from underneath the power line.

Lightning Bugs designed for nocturnal reconnaissance missions were fitted with strobe lights to provide illumination for their cameras. Although the strobe lights at first proved ineffective as photographic tools, they had the unexpected side effect of disturbing enemy pilots and soldiers on the ground. We can imagine that strobing drone sightings by uninformed pilots and soldiers on both sides of the battle line may have contributed a few dramatic stories to the UFO lore.

These aircraft, deception and ECM measures are all between forty and fifty years old, their technology mostly declassified and available to the public, but the principles that they employed then remain much the same today, only considerably more sophisticated. The new generation of UAVs fly longer, faster, lower and slower than the Firebees ever could, and are playing an increasingly prominent role in modern warfare and in furthering the UFO lore.[6]

One dramatic technological development long rumoured to be part of the modern drone's defensive arsenal is the ability to disappear from sight, using what is called active or adaptive camouflage – the drone is now a chameleon, blending into the sky above. The origins of this optical stealth technology date back to the Second World War when the RAF and the US Navy, under the name Project Yehudi,[7] mounted lights on to the wings of bomber planes used at sea. Using dimmers, the lights could be matched to the luminance of the sky, softening the sharp outlines of the planes' wings and fuselage on a bombing approach and reducing the range at which they could be targeted from about twelve miles to under two. The introduction of radar made Yehudi redundant, but the shapes and colour schemes of today's radar-evading drones are carefully optimized to avoid casting shadows and to blend in with the sky. Since the early 1980s various attempts have been made to update the Yehudi concept, combining radar and optical stealth to render an aircraft invisible to machines and humans. Paul Bennewitz told other ufologists that the craft he saw around the Manzano storage area were able to bend light around themselves to become 'invisible'.

As if taunting us with this new capability, Boeing's Bird of Prey, a piloted craft, and the still-secret Advanced Technology Observation Platform (ATOP), probably a drone, both make sly reference to the 'cloaking' abilities of the Klingon 'Bird of Prey' spacecraft in the fictional *Star Trek* universe: a mission patch for the ATOP shows the outline of the Klingon warcraft, through part of which the background of an American flag is visible. It's rumoured that the new generation of optical stealth technology employs a material that can change colour when electrically charged, responding to sensors or fibre-optic cameras on top of the aircraft. Some witnesses of 'black triangles' have described them blending in with the stars overhead, suggesting that the technology may already be in use.

Useful as Blackbirds and Firebees have been in reconnaissance roles, the military's most effective eyes in the sky now fly several miles above the Earth. The launch of Sputnik in 1957 opened up a new frontier in the Cold War: space. The 1959 launch of the joint CIA–Air Force reconnaissance satellite project, codenamed CORONA, saw a dramatic change in the way visual intelligence was gathered and a decreasing reliance on risky manned spyplane missions, even if these have never been entirely phased out. CORONA satellites would photograph targets on the ground miles below then release their film rolls for a fiery descent to Earth, where they would be snatched up mid-fall by tricked-out aircraft. The capsules, shaped like a large metal bucket, or perhaps a misshapen flying saucer, would sometimes miss their targets. One landed in Venezuela on 7 July 1964, sparking a desperate, potentially catastrophic and highly embarrassing CIA 'bucket-hunt' – a true crashed UFO recovery. Similar accidents are on the record; we can assume that many more aren't. CORONA flew its final mission in 1972, under the auspices of the National Reconnaissance Office, whose very existence was a secret until 1992; the project ended after being compromised by the Russians, who had a submarine waiting below the site of a film canister re-entry.

Today the US Air Force, the Army and the Navy all have dedicated Space divisions, like US Air Force Space Command, formed in 1982, and the Space Warfare Center, formed in 1992, and rumours of reusable militarized space planes and shuttles constantly bubble around the black world of aviation enthusiasts. While such aircraft have yet to take on tangible form, the extreme end of the UFO community has morphed their hypothetical existence into entire space fleets, built to counter a possible threat from Out There.

As many of the world's nations launch their own rocket programmes, it will become ever more difficult for the superpowers to hide what they're doing in space. It's getting awfully crowded up there. Just as counter-measures were developed to deceive ground- and air-based radar systems, we can bet that the brightest technical minds in the intelligence world have found ways to mask the launches, orbital paths and re-entries of their space-based hardware. And this, of course, means more UFOs.

PIES IN THE SKY

U-2s, Blackbirds, drones, Stealth Fighters and bombers – all of them are likely to have contributed, directly or indirectly, to the UFO lore of the past sixty years. These are all aircraft that we know about; but what about the ones that never got past the prototype stage, or remained cloaked in the twilight world of the 'black' budget? Many of them, it turns out, were flying saucers.

The Air Force's first top-secret UFO study, *Analysis of Flying Object Incidents in the United States*, internally published in April 1949 and made public only in 1985, includes a handful of flying saucer sightings that pre-date Kenneth Arnold's. A fast-moving flying disc was seen in April 1947 by two employees of the Weather Bureau Station at Richmond, Virginia, while in Oklahoma City RCA engineer Byron Savage described a large, fast-moving disc, 'perfectly round and flat' fly overhead with a whooshing sound.

The report also includes what must be the world's first photograph of a UFO, taken by William Rhoads on 7 July 1947 over Phoenix, Arizona. He also heard a whooshing sound and, thinking it was a jet fighter, rushed outside with his camera. Instead he saw a strange heel-shaped object circling at high speed above him. Rhoads's photographs were printed in a local newspaper and drew the attention of the FBI, who obtained his negatives and declared them to be genuine photographs, though of what, we don't know. Perhaps it was the heel of a shoe, but it's a decent match to Arnold's own initial sketch of what he saw on 24 June, with a rounded front and straight sides meeting at a convex end.

So was there anything flying at the time that could account for these sightings? The answer is a frustrating maybe.

While they seemed a novelty to the newspaper-buying public of 1947, flying discs were nothing new. In 1716 the science journal *Daedalus Hyperboreus* published a diagram of a round-winged flying machine designed by the Swedish scientist and mystic Emanuel Swedenborg, who would himself experience contactee-style transcendental journeys to other planets later in life.

In the mid-1930s, Arup Inc. of northern Indiana successfully developed four models of a heel-shaped, single-winged aircraft, while in Ohio a team under designer Steven Nemeth built the Roundwing, a disc-shaped single-wing on top of a more conventional propeller-driven fuselage. Flying discs also began to grace the covers of science-fiction – or 'scientifiction' – magazines around this time.

By the end of the Second World War, so much of which had been fought in the air, the major powers began to recognize the importance of short landing and take-off capabilities for aircraft. The Army Air Force learned that bombing an airstrip effectively paralysed both outgoing and incoming air traffic, while the Navy sought new approaches to landing and launching aeroplanes from ships. It was the US Navy who commissioned aircraft designer Charles Zimmerman at Chance Vought to develop a short-take-off-and-landing, propeller-driven fighter plane.[8] This became the XF5U-1, the 'Flying Flapjack'. Officially only two aircraft were built and, long over budget and overdue, the project was scrapped in March 1947 when the Navy decided that the future lay in jet propulsion. For inexplicable reasons the Navy ordered the manufacturers to scrap their fully functional, flying-test model, wasting thousands of dollars in materials and research. However, designs exist for an advanced model of the Flapjack, known as the Skimmer, which allegedly had the ability to hover, while rumours of an advanced, jet-propelled variant of the Flapjack persist to this day. If either did exist they remain remarkably well-kept secrets, and, ironically, the best evidence we have for their existence would be the heel-shaped UFOs described by Kenneth Arnold and photographed by William Rhoads.

It's extremely tempting to tie these early flying saucer reports to the existence of one or more prototype Skimmers or Flapjacks. The Navy's insistence that their test model be scrapped might suggest that they were developing something similar to replace it. At least one engineer, Thomas Clair Smith, who would become Vice President of the enormous Woodstream Corporation, claims to have witnessed flights of a Skimmer at the Vought plant in 1946: 'Did the aircraft fly?' he told one reporter. 'You bet, I saw it take off, hover and land.'[9] Smith stated that the Skimmer he saw was tested only at night, but perhaps by 1947 it was venturing into the daylight.

Leon Davidson, the scientist and author of 'ECM+CIA=UFO', was convinced that the earliest saucer sightings were indeed of experimental Navy craft, probably the XF5U-1. Davidson decried the initial US Air Force investigations into the saucer mystery as 'one branch of the Department of Defense . . . investigating public manifestations of a secret activity of other branches'.[10] Certainly if the Navy were flying heel-shaped, hovering aircraft in 1946 and 1947, we can bet that they'd have taken great pleasure in running rings around the Air Force and causing them as much embarrassment as possible. Davidson believed that Navy harrassment could explain the very first Air Force UFO sighting, of a 'thin metallic object' over Muroc AFB on 8 July 1947. Davidson also identified the 'Hell Roarer', a magnesium lighting device that burned at ten-million candlepower, as a source of UFO sightings; it certainly caused a flood of saucer reports when tested over Connecticut in October 1951.

But the Americans weren't the only ones developing unusual aircraft during the 1940s and 1950s, and it's here that the historical record begins to distort and shimmer like tarmac on a desert landing strip.

HAST DU GESEHEN DIE FLUGKREISEL?

Wernher von Braun's V-2 rocket had come startlingly close to turning the tide of the Second World War in Germany's favour. In the war's aftermath all the remaining rockets and parts were quickly snapped up by the Americans, along with von Braun and more than 200 key scientists and engineers as part of Project Paperclip. All hardware and personnel would eventually be filtered through the Air Technical Intelligence division at Wright Patterson AFB in Ohio – later home to the Air Force's UFO investigations. Fifty-eight German aircraft and hundreds of engines and parts made their way there, the V-2 and the Messerschmidt 262 jet fighter being the most celebrated among them.

Numerous other interesting secret weapons are said to have been in development for the Axis's air arsenal. One craft we know existed was the Horten Go 229 Flying Wing, designed by Walter and Reimar Horten and successfully test flown in 1944. The proposed fighter-bomber had a fifty-five-foot wingspan and is thought to have had an extremely small radar signature, making it an ancestor of America's B-2 Stealth Bomber. A prototype Flying Wing was captured by the Americans at the end of the war, along with plans for other Horten aircraft. America had been working unsuccessfully on similar designs since 1944, and in October 1947 began test flights of the enormous, but troubled, 173-foot wingspan Northrop YB-49.

The Russians also grabbed their fair share of German technological booty, leading Russian-piloted aircraft based on Horten designs to become prime flying saucer candidates during the first wave of UFO activity over America. The second drawing made by Kenneth Arnold after his sighting depicts a crescent-shaped craft strikingly similar to the Horten Flying Wing and is quite different from his first drawing, which better resembled the US Navy's Flying Flapjack. Was this transformation significant? Was Arnold asked to reconsider his drawing by anxious Navy admirals?

In March 1950, as Europe experienced its first flush of UFO reports, its popular presses ran two stories from engineers claiming to have worked on flying saucer aircraft during the Second World War. The first was by Professor Giuseppe Belluzzo, a famed Italian engineer who had sat in Mussolini's Senate and written more than fifty books. Belluzzo announced that the flying discs were a type of aerial torpedo and had been under development in Germany and Italy since 1942; his statement appeared in several Italian dailies and was picked up by the Associated Press on 24 March. Belluzzo was quoted as saying: 'There is nothing supernatural or Martian about flying discs . . . they are simply rational application of recent technique.'

Just a few days later, on 30 March, as part of an article about flying saucers over Europe, Germany's *Der Spiegel* ran an interview with Rudolf Shriever, a driver for the US Army at Bremerhaven on Germany's north coast. Schriever, described as a former Luftwaffe Captain and Prague University graduate, expanded on Belluzzo's claims, stating that he had worked on the Nazi flying disc project from 1942, and that a working model had flown in 1945. Schriever's design featured a central crew cabin around which a circular rim rotated at great speed, powered by jet turbines underneath it; the forty-nine-foot craft was capable of fantastic acceleration and could reach speeds of up to 1,200 m.p.h. As other publications picked up Schriever's story, he went further, revealing that a flying prototype had either been destroyed or fallen into Russian hands, and that his own plans for the craft had been stolen from his workshop after the war. Schriever was convinced that the Russians or the Americans were flying craft based upon his designs.

Adding to the mystery, he reportedly died in a car accident in 1951, a year after going public with his story.

Schriever's claims are vague, but some have been corroborated elsewhere. The February 1987 issue of German aviation magazine *Flugzeug* ran an article by a German pilot who said that, in August or September 1943, he and other trainee pilots at Praha-Kbely airfield, just north-east of Prague in the Czech Republic, had witnessed the test flight of a twenty-foot-diameter flying saucer. The aluminium craft, which had a central cabin and a rotating outer rim, sat on four legs and rose a few feet off the ground – not quite as impressive as Shriever's saucer, but suggestive none the less of a genuine programme.

That wasn't all we would hear about Schriever's *Flugkreisel* (rotary flyer or flying top). On 7 June 1952, coincidentally the same month that Major Donald Keyhoe's pro-ET *The Flying Saucers are Real* hit American bookstands, *France Soir* newspaper was running more tales of Nazi flying saucers. This time the whistleblower was Dr Richard Miethe, who claimed to have been one of the German disc's principal designers, alongside Shriever. Miethe also had photographs of its test flight, which were published in Italy's *Il Tempo* newspaper the following September. Revealing only an indistinct blob, the images could be of almost anything, including a flying saucer.

In the *France Soir* article, Miethe describes a 'supersonic helicopter', the V-7, which he concedes could, at a distance, 'resemble the saucer of a set of tableware' and 'has the exact shape of an Olympic discus, an immense metal disc of circular form, with a diameter of approximately 42 metres'. According to Miethe, the V-7 flew using twelve powerful jet engines that rotated around a central shaft and, thanks to a cunning 'compression system', produced neither flames nor smoke. Miethe claims that the V-7 was intended to go into mass production towards the end of the war, and that the Russians managed to get hold of parts of one of the craft, as well as plans and three engineers.

These three accounts formed the basis of a rich, ongoing Nazi flying disc lore that now incorporates tales of surviving Nazi communities in Antarctica and Latin America and lurid, time-travelling, *ubermensch* escapades. During the 1970s the myths were adopted and expanded upon by neo-Nazi propagandists who hoped to use the existing popularity of the UFO subject to draw readers into a powerful and engaging, if rather unlikely, fantasy of Nazi supremacy.[11]

SOVIET SAUCERS

There's little evidence for Schriever or Miethe's claims beyond this initial flurry of press reports. But what about their statements that Russia was flying its own discs?

The CIA's 1953 Robertson Panel report on UFOs expressed concern that the Soviet press had run no stories about flying saucers: 'The general absence of Russian propaganda based on a subject with so many obvious possibilities for exploitation might indicate a possible Russian official policy.' Did the Russians deliberately keep silent on the flying saucer issue because they, too, were being visited by the mystery craft or, as many Americans suspected at the time, because the flying saucers were their own?

Three tales from the 1950s may point to a working Russian disc project. The first, reported in the *News Dispatch* on 9 July 1952, involved a forty-eight-year-old former mayor, Oscar Linke, and his daughter Gabriel, whose sighting, near Hasselbach, took place two years before on 17 June 1950. The Linkes had found themselves just fifty feet away from two human figures dressed in shimmering, metallic heavy garments, 'like people wear in polar regions'; they also wore flashing lights on their chests. Thirty feet from these figures was a classic flying saucer, between forty and fifty feet in diameter, with two rows of small foot-wide holes surrounding it at about eighteen-inch intervals. A ten-foot-high 'conning tower' stuck out of its centre. When Gabriel called out to her father the figures became alarmed and clambered inside. The outer rim began to rotate, and the holes glowed with flames, like a jet engine. The disc rose vertically up the shaft of the central tower until it looked like a mushroom, then the whole craft moved parallel to the ground before taking off with a loud whistling, roaring sound. Two other witnesses saw the disc fly by. 'I thought it was a new Russian war machine,' Linke was quoted as saying. 'I was terrified, for the Soviets do not like one to know about their goings-on.'[12]

The next rather charming tale appeared in France's *Le Quotidien de la Haute-Loire* newspaper on 24 October 1954. Czech factory worker Louis Ujvari was on his way to work in St Rémy in Central France at about 2.30 a.m. when he ran into a man of average height, dressed in a pilot's flight suit and wielding a revolver. At first the man tried to communicate with Ujvari in a language that he didn't understand, so the Czech suggested they speak in Russian.

The pilot then asked in Russian, 'Am I in Spain or Italy?'

When told that he was in France, the pilot first asked how far it was to Germany, then what time it was.

'It is 2.30 a.m.,' Ujvari told him.

'You lie,' the pilot replied. 'It is 4 a.m.'.

Following this oddly pointless exchange the pilot turned around and walked towards his craft parked at the side of the road. It was shaped like two inverted saucers with a dome on top, from which protruded an antenna. A bright light came on, then the craft rose up with a high-pitched whine like 'a sewing machine' and disappeared into the night.

We might comfortably dismiss these two stories as apocryphal, or as having been fabricated by the CIA to point the finger of saucer suspicion at the Soviets (we shouldn't underestimate the intricate lengths to which both sides would go to deceive each other). But our final account, from 4 October 1955, is much harder to ignore. Its source was Senator Richard Russell, head of the US Armed Services Committee and one of the nation's most knowledgeable experts on defence matters. The Air Force Intelligence report that details his sighting, dated 14 October 1955, was classified 'top-secret' until 1985. It describes how Russell and two men, his military aide Colonel E. Hathaway and a 'businessman' whose name is blacked out in the report, were travelling by train near Baku in present-day Azerbaijan. It was twilight, just past 7 p.m. and the skies were clear. Russell was alone in his carriage when he noticed an odd yellow-green light rise vertically into the sky about a mile away before passing right over the train. Russell called for his companions and described what he saw. As he did so, another light appeared in the same location and again flew overhead.

'We've been told for years that there isn't such a thing but all of us saw it,' Colonel Hathaway told his Air Force interviewer ten days later. 'The aircraft was circular [and] resembled a flying saucer . . . revolving clockwise or to the right . . . There were two lights towards the inside of the disc which remained stationary as the outer surface went around.' The anonymous businessman described seeing a 'slight dome on top' and a white light.

Once Russell and his team had returned to the US, the CIA and the FBI both got in on the act. The CIA tried hard to dismiss Russell and Hathaway's sighting as a misidentification of a missile launch or a jet fighter, concluding that 'even if the present sightings by Russell are confirmed, it should not be inferred that these unconventional aircraft have actually been flying around the US'.

Did the Senator and his team really see flying saucers? Given Russell's status as one of the major players in the American military establishment, it's hard to imagine that anything on his journey would have happened by accident: so were they actual flying saucers, or more conventional craft mistaken for, or even disguised as, them? We might imagine that if the Russians really were flying advanced vertical-take-off-and-landing (VTOL) aircraft they wouldn't want the US to know about it. Conversely, if they weren't flying them, they might want to alarm the Americans into thinking that they were.

The CIA were as perplexed as you probably are, but they also knew that a Russian flying saucer wasn't that unlikely a proposition, because America was building its own – codenamed Project Y.

FIFTEEN
THE SECRET WEAPON

'Just keep your eyes on it and when you see something come out of it, that will be Jesus'

UFO witness Vickie Landrum to Colby Landrum, New Caney, Texas, 29 December 1980

The history of man-made flying saucers is a tangled one. Once I'd managed to piece together the fragments I was left with one burning question: why, if disc-shaped aircraft had been flying since the 1930s, did the flying saucer acquire such a magical, otherworld aura?

The Swiss psychologist Carl Jung, who was a close friend of CIA director Allen Dulles, became fascinated with flying saucers late in life, and his 1959 book on the subject, *Flying Saucers: A Modern Myth of Things Seen in the Sky*, remains a classic. Jung attempted to explain the disc's power by equating it with the mandala, an image of totality, wholeness and perfection – properties otherwise found only in god. And god, the mythical sage Hermes Trismegistus tells us, is a circle whose circumference is everywhere and whose centre is nowhere, a paradoxical analogy fitting to the UFO problem. It's not just the circular form of the craft that lends them their mystery, it's also what Jung calls the 'impossibility'[1] of so many of the reports, an inherently dream-like and illogical nature that suggests that at least part of their landing gear resides permanently in the imagination.

We might, however, wonder if there wasn't some deliberation behind the development of flying saucer's mystique, which made it as much of a psychological weapon as it was a nuts-and-bolts aircraft. This makes a lot more sense when we realize that by the early 1950s America's own flying saucer project was well under way.

On 11 February 1953 the *Toronto Star* announced that flying saucers were no longer a fantasy – Avro Canada was building one at its Malton airfield. The vertical-take-off-and-landing (VTOL) craft was, according to the *Star*, capable of hovering at a near standstill or shooting away at 1,500 m.p.h. The story was mostly true; Avro was working on the craft, known as Project Y, or the Avro Ace because of its spade-like delta shape, but it existed only as a mock-up and was nowhere near to getting off the ground.

British aircraft designer John Frost, responsible for the RAF's famous De Havilland Vampire, had developed the Ace to meet increased demand for a VTOL aircraft. Like the rumoured German flying discs, the Ace was to use a flat radial-flow-turbine engine, nicknamed the 'pancake', which rotated around a central axis and pushed air out along the aircraft's rim. Like many in the aviation world, Frost was convinced that the Russians were already flying disc-shaped aircraft based on German designs, and is said to have sought advice on the Ace's design from engineers who worked on Heinkel-BMW's *Flugkreisel*.

By 1954 the Canadian government was running out of both patience and money and Project Y was at risk of closure. Enter Lieutenant General Donald Putt of the US Air Force Research and Development (R&D) department. Putt offered Frost $200,000 to continue his work and then, in 1955 – perhaps spurred on by Senator Russell's sighting the previous October – another $750,000.[2] Out of the Air Force money came Project Y2, incorporating plans for two saucers known as Silverbug and

Ladybird. Tellingly perhaps, it was around this time, in October 1954, that the US Department of Defense's form letter about flying saucers dropped a paragraph denying that the 'unidentified aerial phenomena' were a 'secret weapon, missile or aircraft developed by the United States'.[3]

Intended as a single-pilot interceptor, the more advanced Ladybird was expected to reach astounding speeds of Mach 3.5 (more than 2,600 m.p.h.), to fly at 80,000 feet and to be able to reach 70,000 feet from a hover in an improbable four minutes twelve seconds. The discs would achieve these speeds by exploiting the coanda effect, in which the air flowing around the rim of a curved surface generates additional lift. A number of roles were proposed for the Y2, including launching it from Navy ships and submarines, using smaller versions as unmanned flying bombs, and fitting the piloted version with a hardened rim, allowing it to 'slice' its way through enemy aircraft – not a job that many pilots would have relished.

The most advanced version of the Frost disc, the Avro MX-1794, incorporated more powerful turbojet engines and reached the mock-up stage in 1956. Wind-tunnel tests took place at Wright Patterson, no doubt contributing to rumours about flying saucers at the base, and then at the NACA – precursor to NASA – facility at Ames, California. Things were looking good for Project 1794: at a Navy presentation in October 1956, Avro announced that a flying prototype would be completed in January 1957 . . . and that's the last that anyone heard of it.

So what happened to America's flying saucer? Aviation historians Bill Rose and Tony Buttler see the confusing use of multiple project names for essentially the same aircraft as deliberate obfuscation, and suggest that MX-1794 went 'black' in its final stages. The authors claim to have seen US documents from 1959 discussing an ongoing flying saucer development programme, with Lockheed's famous Skunkworks, home of the U-2 and Stealth planes, as a likely location.

As a bittersweet postscript to the Silverbug story, in 1958, just as the MX-1794 vanished from sight, Avro announced a new project, the VZ-9AV, best known as the Avrocar, an eighteen-foot-wide, three-foot high, single-pilot flying saucer. Intended as a hovering jeep for the Army, the Avrocar turned out to be a juddery, unstable and ultimately useless dud whose only role seemed to be providing comic turns in newsreels – a deliberate distraction, some say, from the real and top-secret MX-1794. Until the Silverbug papers were declassified in 1995, the Avrocar was all that the general public knew of the Avro disc project, the result being that a home-grown saucer programme was never considered a serious possibility.

The US Air Force ceased operations with Avro Canada in 1961, and its Malton plant shut down for good the following year, but that wasn't the end of its flying saucer dream. Codenamed Pye Wacket after the magical cat in the 1958 film *Bell, Book and Candle*, the Convair Lenticular Defense Missile was a small radio-controlled discus, about five feet across, that was to be launched from the short-lived B-70 Valkyrie bomber. Pye Wacket would be capable of manoeuvres impossible for any piloted aircraft, which was just as well as it was intended to detonate on impact at speeds of up to Mach 7 (about 5,000 m.p.h.). The platform reached wind-tunnel testing stage before being cancelled in 1961, and photographs of small, sleek saucer models still exist. As with the MX-1794, it's possible that the Lenticular Defense Missile went black: rumours persist that full-scale models were tested at White Sands, and something like Pye Wacket would account for the many sightings of smaller, discus-shaped UFOs over the years.

No discussion of man-made UFOs can ignore the enigma of Thomas Townsend Brown, an Ohio-born engineer and founder member of NICAP, for many years America's largest UFO study group. A natural-born tinkerer, in 1921 Brown discovered the propulsion effects generated by ionized electrons; he was just sixteen. The Biefeld–Brown effect, usually called electrohydrodynamics (EHD), is the principle behind today's 'lifters', light airframes that use electricity as their source of propulsion, a principle that may one day play a role in Deep Space exploration.

Much of Brown's career is shrouded in secrecy – during the 1930s and 1940s he worked for the Army and Navy on a number of still-classified projects, probably involving radar and satellite communications. By the late 1950s Brown was said to be working on anti-gravity research using model flying saucers, part of a wave of interest in the subject reported in newspapers and popular magazines like *Mechanix Illustrated*.[4] Antigravity research was alleged to be taking place at fourteen sites around the US, including the Aeronautical (later Aerospace) Research Laboratories at Wright Patterson, where the Avro disc was being tested. Some in the UFO community believe that an antigravity breakthrough was made during this period; more likely is that Brown's ideas led to small advances in countering gravity's pull, for example in the use of electrical fields to reduce drag on aircraft wings, a system reportedly employed today by the huge B-2 Stealth Bomber.

There's one more area that we need to consider in the search for man-made UFOs. As part of its enthusiasm for all things atomic, in 1946 the US Air Force founded Nuclear Energy for the Propulsion of Aircraft (NEPA). Actual test flights didn't begin until 1955 when, over a period of two years, a B-36 bomber fitted with a twelve-tonne lead and rubber shield underwent forty-seven test flights over Texas and New Mexico with an air- and water-cooled nuclear reactor on board. Plans were made for an atomic aircraft, the X-6, and prototype engines were built, but the entire programme was scrapped in 1961. But did it really end? A handful of UFO encounters have left witnesses with what appear to be the symptoms of radiation exposure, most famously the Cash–Landrum incident of 29 December 1980.[5]

This is by no means a complete history of disc-shaped aircraft. Many aircraft designers were inspired to experiment with the form by the constant flow of UFO reports, and rumours of saucers held in hangars at American airbases persist to this day. Popular variations on the tale include large saucers spotted in the scrap yard at MacDill AFB in Florida in the 1960s and whispers of a trio of antigravity Alien Reproduction Vehicles (ARVs), known as the Three Bears, being back-engineered at a secret location in the US during the 1980s.

UFOS: UNFUNDED OPPORTUNITIES

Either by accident or design, during the 1950s the flying saucer became associated with an intelligence and technology that was not of this Earth; and in the post-war battle for global supremacy, the promise of advanced alien technology would be a powerful weapon in the psychological warfare arena. Contrary to the UFO community's cries that the US government was silencing the truth about UFOs, as the Cold War ratcheted up to ever greater degrees of intensity, people such as Silas Newton, Olavo Fontes and, later, Paul Bennewitz were encouraged to believe that the US military was in possession of advanced extraterrestrial technologies hundreds, if not thousands, of years in advance of British technology. These ET craft could travel at impossible speeds then stop and hover on a pinhead; they could be silent; they could be invisible – they may as well have been invincible. And they belonged to America.

In the years following the Bennewitz affair these rumours grew louder than ever before. In the February 1988 issue of *Gung Ho* ('The Magazine for the International Military Man'), the pseudonymous Al Frickey[6] was one of the first to draw attention to stories of ET craft being flown at Area 51. The Nevada base, technically part of the Nellis AFB Military Operations Area, is now a household name in many circles, but back then its very existence was highly classified. The article's title, 'Stealth – and Beyond – a look at Aurora and Some "Unfunded Opportunities" (UFO)', referenced a comment made by Ben Rich, director of the Skunkworks, Lockheed Martin's Advanced Development Program, home of the U-2, the SR-71 Blackbird, the Stealth Fighter and the B-2 bomber. When asked for his opinion about UFOs, Rich is said to have replied, 'UFOs represent Unfunded Opportunities'. It's a multi-layered comment that cuts many ways: does he mean that Skunkworks developed flying saucers like the Avro MX-1794 but the money ran out? Or are the UFOs that people see prototype aircraft that haven't yet made it to the development stage?

Speculating about what might be going on out at Area 51, especially in its Alien Technology Center ('Do you think they are studying Mexicans?' asks the editor), the *Gung Ho* article quotes an anonymous Lockheed engineer: 'Let's just put it this way. We have things flying in the Nevada desert that would make George Lucas drool.' A nameless Air Force officer tells Frickey: 'We are flight-testing vehicles that defy description. To compare them conceptually to the SR-71 would be like comparing Leonardo da Vinci's parachute design to the space shuttle.' Frickey also hints at rumours of 'force-field technology, gravity-drive systems, and "flying saucer" designs'. What super weapons were they building out there at Area 51?

In his Second World War memoir *Most Secret War*, British radar pioneer Reginald V. Jones describes the fear that a 1939 speech by Adolf Hitler induced in British intelligence circles. In the Foreign Office translation of his speech, Hitler bragged that the Germans had a 'secret weapon against which no defence would avail'. In a follow-up pronouncement, Hitler said that this weapon would deprive its victims of 'sight and hearing'.

The Secret Intelligence Service called upon Jones to assess what this secret weapon might be, and he summarized the possibilities in a report:

> Apart from the more fantastic rumours such as those concerned with machines for generating earthquakes and gases which cause everyone within two miles to burst, there are a number of weapons . . . of which some must be considered seriously. They include:
>
> Bacterial warfare
> New gases

Flame weapons
Gliding bombs, aerial torpedoes and pilotless aircraft
Long-range guns and rockets
New torpedoes, mines and submarines
Death rays, engine-stopping rays and magnetic guns.[7]

The truth about Hitler's speech ended up being rather more prosaic. The German for weapon is *waffe* and once Jones had the speech retranslated it became clear that Hitler was referring to the *Luftwaffe*, his air force. And the mysterious force that would make its victims blind and deaf? This was merely an awkward interpretation of what we would call 'thunderstruck'.

Bluff and bluster are critical weapons in any nation's armoury. Whether in war or in peacetime, if an adversary or a potential adversary claims to have its hands on advanced technology, then you can't afford to ignore it. At the very least you will expend considerable man hours and money trying to assess whether or not the claims are true. You may waste further resources trying to build one yourself. Meanwhile, the effect of such threats on civilian and military morale can be just as devastating as the alleged weapon itself.

During the early years of the UFO phenomenon, it was feared that the ghost rockets, flying discs, green fireballs and other mystery objects might represent advanced Soviet technologies. Following the death of saucer-chasing pilot Thomas Mantell in 1948, the situation became even more dire: if US Air Force pilots were too afraid to confront unknown aircraft because they might end up like Mantell, then all hope was lost. What if one day the flying saucers did prove to be Soviet aircraft – who would dare to attack them? This couldn't be allowed to happen and was one reason why the UFO panic had to be stopped, with tighter controls on military reporting of sightings, and the discouragement of civilian excitement about the saucers. Conversely, by encouraging *foreign* nations to believe that the flying saucers were invincible, and in US hands, they might discourage enemy pilots from attempting to bring down their U-2s or reconnaissance balloons.

STAR WARS

The super weapon ploy is far from just speculation. Consider this analogous scenario, played out in 1983 while Paul Bennewitz was at the height of his mania.

At the time, the possibility of Mutually Assured Destruction had both East and West in the grip of mortal fear. The CIA projected that at the current rate of expansion, the USSR would have 21,000 nuclear warheads capable of reaching the USA by the decade's end. A single missile strike would be catastrophic, making the CIA's vision nothing short of apocalyptic. Something had to be done to defuse the situation.

That March, in a nationally televised speech, US President Ronald Reagan announced that America was soon to become impregnable to Soviet missiles. The American people would 'live secure in the knowledge that their security did not rest on the threat of US retaliation to deter a Soviet attack, knowing that we could intercept and destroy strategic ballistic missiles before they reached our own soil and that of our allies'.

Reagan's super weapon was the Strategic Defense Initiative (SDI), known as Star Wars; it would, in theory, create a 'leak-proof Astrodome' over the heads of America and its allies by shooting down enemy missiles before they inflicted damage on US soil. As a bonus, if the US could rest easy under its bulletproof glass ceiling then they could also, if pushed, launch a first nuclear strike against the USSR. Congress was sceptical about SDI, but without their approval there would be no money to develop it. So, to persuade Congress to loosen its purse strings, and to convince the Russians that Star Wars was deadly serious, the Pentagon devised a simple, but brilliant deception.

Over four demonstrations, missiles playing the role of Soviet ICBMs were launched from Vandenberg Air Force Base in

California, while SDI interceptor missiles were launched from the South Pacific. The first three tests failed, but the fourth SDI missile made a direct hit. Footage of the successful intercept was broadcast on TV news all over the world, trumpeting the warning that America's military superiority should never be questioned. An impressed Congress pumped a colossal $35 billion dollars into SDI, and the Pentagon's plans for the programme grew ever more advanced. Russia, already facing economic disaster, couldn't compete. Or they wouldn't have been able to, if the the whole thing hadn't been a tremendous piece of stage management.

SDI was a hoax; the demonstration intercepts had been simple parlour tricks. The dummy Soviet ICBMs carried bombs that could be detonated when the SDI missiles got close enough to make the effect look convincing. The problem was that the first three interceptors missed their targets by such a distance that detonating them would have looked ridiculous. For the fourth, 'successful' test, the incoming missile was artificially heated and fitted with a radar beacon, creating a bigger target for the inept interceptor, which was now also armed with heat-seeking sensors, just to be on the safe side. The target missile was as close to a sitting duck as something hurtling through the air at 9,000 m.p.h. can be.

SDI was one of the more spectacular demonstrations of Pentagon deception, and one that, thankfully, helped to defuse Cold War tensions – unlike, say, the 1964 attacks on US naval vessels in the Gulf of Tonkin that took America into Vietnam, or Saddam Hussein's non-existent weapons of mass destruction.

Is the extraterrestrial super weapon just another miltary myth? Have our own earthbound technologies become so advanced as to be indistinguishable from ET craft? Or are the Pentagon's UFOs really nothing but weapons of mass deception?

SIXTEEN
A MATTER OF POLICY

'The public's right to know is the Russians' right to know. The Russians read our newspapers and magazines and technical journals very carefully indeed.

CIA director Richard Helms, interviewed by David Frost in 1978

The blueprints for the UFO myth were complex and several pieces were still missing, but an outline was now taking shape. When the first wave of flying saucer reports hit US newspapers in the summer of 1947, those in the know at Wright Patterson were aware that a kernel of truth was buried among the mis-identifications and the press hysteria. There was nothing futuristic about flying saucers – they were just another way of building aircraft, one that had been around for decades. Air Technical Intelligence's R&D team knew that the Germans had at least thought about building fixed-wing and circular aircraft; they knew that the Russians might, conceivably, have already built one, and they probably knew that the US Navy had got their own Flying Flapjack into the air.

By the early 1950s the US Air Force and the CIA had their own working UFO to hide, the U-2, as well as new saucer aircraft designs like those being developed at Avro. A decade later the

Oxcart and the Bluebird were careening through near-space escorted by flocks of unmanned drones and, another ten years after that, the Stealth planes were making their discreet debuts. If there were real alien spacecraft up there, then they would be difficult to spot in these already crowded skies. These secret Cold War technologies formed the skeleton of the UFO myth. The finer, fleshier details had been filled in by the imaginations of the people on the ground, encouraged and embellished by AFOSI, the CIA and others in the alphabet soup of intelligence organizations.

The story John and I had pieced together brought us as far as Bill Moore's public confession in 1989. But the UFOs, and the stories people told about them, kept on coming: a UFO spotted over Chicago's O'Hare Airport on 7 November 2006 generated a flurry of wild media excitement, as did a spectacular series of sightings at Stephenville, Texas in January 2008. And then there was Serpo. Although we suspected that this was a dirty tricks operation, we were a long way from proving it. We needed to find evidence that the Mirage Men were still active, that running UFO disinformation campaigns was not just a thing of the past, but business as usual for the intelligence agencies.

It was at this point that the tables were turned and I became a target. It began with an anonymous email forwarded to me by an apparently concerned Rick Doty:

> [name removed], a senior DIA official, told me, point blank, that a name check revealed that Mark Pilkington, a British citizen and Independent Film-Maker, is an MI6 source. He was used to spy on other film-makers and is currently a source of information to British Intelligence on Crop Circles [sic]. He is paid by British Intelligence and has been since October 1998 . . . If I was you, I'd be careful of my 'circle of friends'.

Soon afterwards another contact in the UFO field got in touch: 'Just to let you know that I have just received a phone call, from someone I won't name, warning me you're British Intelligence . . . It's doing the USA circuit big time, evidently.'

The UFO community's relationship with authority is conflicted at the best of times. They believe that their governments are hiding the truth about UFOs, yet they await with fervour the day that the same governments reveal all in that apocalyptic moment they call 'disclosure'. At this point, the scales will fall from the world's eyes, the extraterrestrials will descend to Earth in a light show to make *Close Encounters of the Third Kind* look like a cheap home movie and, most importantly, the ufologists will be proved to have been right all along.

When it comes to military and intelligence 'insiders', things become more complex still. The UFO scene is overrun with whistleblowers who regale us with tales of underground bases and intergalactic pacts while waving impressive-looking documents around as evidence. To the believers these people are clearly on the side of ufology and are to be welcomed. That the role of AFOSI and other agencies in distributing this same material has been public knowledge for twenty years seems not to have sunk in. Meanwhile, those insiders who suggest that the UFO phenomenon is a complex brew of security, secrecy, psychology, sociology, politics and folklore, perhaps driven by rare but genuinely anomalous events, are obviously part of the cover-up.

In my own case, fingering me as an MI6 agent could have served two purposes. It raised suspicions about me within the civilian UFO community, while scaring off anyone with military or intelligence connections who might otherwise have been inclined to speak to me. The world's intelligence agencies, while they often co-operate, also remain wary of each other. Passing sensitive information to a 'foreign national', especially one with intelligence connections, can land you in extremely deep water. At the same time, the idea that Rick, or anybody else, believed that MI6 would send an agent into the American UFO arena might suggest that there really is something to the UFO story – even if it's ultimately a smokescreen for something else. But perhaps that was what *they* wanted me to think.

Or perhaps the MI6 claim had originated with Rick himself. John and I had achieved what we thought would be impossible, an interview with Rick Doty. More than that, we had become friends, of a sort, with a man we thought we would never meet. But we had only *his* side of the story, and it didn't take an intelligence specialist to recognize that some of the tales he had told us were tall enough to give a skydiving Elvis vertigo. We liked Rick, but we also knew that befriending people and gaining their trust was once his job – he'd told us as much himself. Parts of his story rang true, but plenty more didn't. So were we just the latest victims in Rick's long game?

One interpretation of UFO history holds that Bill Moore and Rick Doty conspired to create the MJ-12 papers as a hoax, perhaps to make money, perhaps just to pull the wool over the eyes of the increasingly hapless and confused UFO community. This didn't wash with us. It was clear that the Bennewitz affair, and the subsequent release of the MJ-12 papers, were elements of a larger disinformation programme sanctioned and conducted by AFOSI. The investigation of Lee Graham at Aerojet and the later FBI probe of the MJ-12 documents clearly pointed to that agency's involvement. Bob Emenegger's engagement with Colonel William Coleman in the early 1970s also demonstrated that AFOSI had been running rings around the UFO scene for years, if not decades.

If John and I could unearth some more recent examples of AFOSI UFO operations, then we could demonstrate that the campaign against Paul Bennewitz wasn't just a one-off but was standard technique, even a matter of policy. We also knew that our own investigations were now making waves among the Mirage Men. Time was running out.

WHAT'S THE FREQUENCY DENNIS?

Roswell, New Mexico is an unremarkable place; flat, drab and rather run down, it's unlikely to win any beauty pageants. Nor would you expect it to win many popularity contests: it's deep inside ranching country, and beyond housing one of the largest mozzarella production plants in the world, it doesn't have a whole lot else going for it. But cities, like people, can't always control their destinies, and Roswell's is now implicated in an event so startling that it can never live up to the expectations placed upon it. And, even if the Roswell Incident did happen as the mythographers describe, nobody's quite sure where: at least three locations compete for the title of official UFO Crash Site.

The confusion over what took place in the summer of 1947, and where, hasn't stopped Roswell becoming the UFO capital of America. Everything here is saucer-shaped, including the sign for the local bank, while the streetlights are topped with tatty plastic alien faces. There are plenty of motels in the city, presumably to welcome visitors to the annual Roswell UFO Festival and Parade (Jefferson Starship recently headlined), and a lot of fast-food joints, many of which were hiring. Young people were notable by their absence: the waitresses in the hip Roswell Cover Up Cafe were refreshingly mature in years. There's also a lot of architectural dereliction on display – the housing out towards the airport, the former Army Air Field where the UFO crash debris was stored, having the appearance of slum dwellings.

Not everyone is happy with the town's intergalactic celebrity status. The receptionist I spoke to at a Christian chiropractor's office complained vocally about what she called 'all that UFO nonsense'. Born just a few days after the alleged crash, she'd heard nothing about it until the early 1990s. Now visitors rarely talk to her about anything else. If pressed she admits that there probably was a military cover-up of some kind, though it had nothing to do with aliens or UFOs – but don't tell that to the folks down at the Roswell UFO Museum.

Located in what used to be the main-street cinema, the International UFO Museum and Research Center is at least partly to blame for stoking up the city-wide UFO fever. The museum first opened its doors in 1991, under the directorship of local developer Max Littell and two key figures in the Roswell UFO story. One was Walter Haut, who penned the press release announcing the Army Air Force's capture of a flying disc. The other was Glenn Dennis, a local mortician who claimed to have been asked to prepare small, hermetically sealed caskets for whatever had died on board the crashed craft.

UFO researchers have since picked Dennis's story to pieces, while Haut's role in the saga seems to have become more dramatic after he got involved with the UFO Museum. What started out with the incontestable fact of his writing a press statement became, in an affidavit released after his death in 2007, the more dubious one of his actually seeing dead aliens and a crashed spacecraft. During later life Haut's statements on the idea of ETs on Earth were unambiguously ambivalent, so this posthumous *volte face* is a little puzzling. The more pragmatic among us might view it as a matter of business rather than as a matter of history.

Nevertheless, the museum itself is well worth a visit. Highlights include a model kit diorama of a Nazi-era flying saucer and, lying in a mocked-up hospital scenario, the prosthetic alien from the 1994 TV drama *Roswell*. The dummy, which Haut named Junior, occasionally appears in closely cropped photographs purporting to be the real live thing: a prop from a fictional film based on an ambiguous interpretation of a historical event has become evidence that this event occurred as described in the film. This epistemological pretzel is a perfect metaphor for the whole UFO experience.

Back in May 1997 the museum and Roswell itself were preparing for what would be the city's finest hour: the fiftieth anniversary of the Roswell Incident. That weekend, hundreds of news crews and journalists, along with thousands of tourists, made Roswell the centre of the UFO universe. The previous year Dennis Balthaser, a newly retired structural engineer with a long-standing interest in

UFOs, had moved to Roswell and begun working at the museum. A smart, friendly and sincere man in late middle age, with a neatly trimmed white beard and a bald pate, Balthaser has since become a prominent member of the UFO community both in Roswell – where he lives with his wife Debbie and their nine pet turtles – and on the national conference stage. While he fell out with the museum's directors a few years ago, he's still involved with the annual UFO festival, but he has never since experienced anything quite like the madness of 1997.

One day in late May of that year a nervous-sounding man telephoned the museum with a story to tell. His father was dying of cancer in a rest home and had only months to live. On his last visit, his dad had given him a small box containing a piece of metal about the size of a dollar coin (three inches in diameter). When crumpled the metal instantly returned to its original form, like intelligent tin foil. When the son asked what it was, his father told him that in 1947 he had been stationed as a military policeman at Roswell Army Air Field. Not only had he helped to clean up the UFO debris field, where he had found the small fragment of the shattered craft, but he had also seen a small child-like alien walking into the military hospital.

The son seemed sincere and, while this wasn't the first time someone had offered the museum a deathbed confession, or even a piece of the Roswell UFO, Dennis Balthaser reckoned he would make the drive out to meet the man at his home in Oklahoma.

During their final conversation before Dennis left, the son mentioned that he'd told a family friend, a retired colonel who had worked in intelligence, about the metal and his plans to hand it to Dennis. The colonel had then told him a few things that he knew about UFOs and ETs. The son also mentioned that he had left the metal fragment at a friend's house for safe-keeping. Both statements made Dennis a little uneasy, but he said nothing, assuming that the son knew what he was doing, and ten days later he set out on his journey.

Once settled in his Oklahoma motel, Dennis rang the son's number. A woman answered and said that the son had left on emergency business and would be back the following day. This seemed a little odd to Dennis, so he checked the phone book under the son's name. There was no entry for him. This wasn't looking good. Dennis checked the address he'd been given and drove over there, only to find a trailer in a trailer park. Hardly where you'd expect an attorney to be living but, thought Dennis, who was he to judge?

The next morning Dennis rang the son three times, getting only an answering machine message each time. He left his motel room number and asked to be called back. He was starting to smell a rat.

In the middle of the afternoon the phone rang. A woman identifying herself as Christie told Dennis that she and a man named Ed would meet him at a Denny's restaurant at 7 p.m. Ed would be wearing a dark suit, she would be in a light-green dress. She requested that if Dennis got there first he find them a non-smoking table.

Dennis made it on time to Denny's. At 7 p.m. exactly a man and woman walked in, dressed precisely as they said they would be, and introduced themselves as Special Agents from Dallas AFOSI. They told Dennis that he would not be meeting the man he had originally come to see; they had visited the son the previous day and picked up the UFO metal, which behaved just as he had described. They also told him that the Roswell Museum's phones were tapped and that they'd found out about his trip five days before he had set off.

Ed then told Dennis lots of strange things: that the Roswell Incident had happened; that the ETs were real, they were here and that the government wanted to tell the world but was still trying to work out the best way to do it. He said that following the Roswell crash the ETs had left a homing beacon transmitting in the desert, and that it had only been discovered by the US government in the 1960s. It sent out a transmission every seventy-three hours and could be picked up at a certain frequency on a CB radio. Ed said that he was one of about fifty agents specializing in UFO matters, and that his boss had been shot dead by the CIA the previous week – why, it was unclear. He said that there were three AFOSI agents and one CIA operative undercover in Roswell, and that Dennis and his colleagues at the museum should be careful about what they said and to whom. He also said that something big was going to happen there soon. Dennis didn't think he was referring to the fiftieth anniversary celebrations.

Three and a half hours later, Ed and Christie left a very confused and bemused Dennis to finish his coffee and ice cream and return to his motel, unsure what to make of it all. About the only thing he knew for certain was that he wouldn't be taking the metal fragment home with him.

Back in Roswell a few days later, Dennis tried one last time to call the son. This time he got through, but was immediately handed over to another AFOSI agent, evidently higher up the food chain, given his curt and businesslike manner. After discussing the situation, the agent identified himself as Charles, and told Dennis that he was from the AFOSI detachment at Langley Air Force Base in Virginia. Charles said that they would be keeping the metal fragment and hoped to be able to identify it, something that they had a good track record in doing. Their conversation ended with Charles insisting that the American public's tax dollars were being well spent when it came to the UFO situation, and that was the last Dennis heard from him. The following September, Dennis tried once more to contact his man in Oklahoma, but a suspicious voice told him that he had a wrong number, and that was the end of that.

There are three ways to read this story.

The first and least convincing explanation is that Dennis Balthaser made up the whole thing. It's true that it's an exciting dramatic narrative that places Dennis at the centre of a cloak-and-lasers intrigue, but Dennis has been telling the same story publicly for several years now without being contradicted by any of his former museum colleagues, some of whom he has, also quite publicly, fallen out with.

The second possibility is that, swept up in the media fracas about Roswell that was brewing at the time, somebody decided to play an elaborate prank on the Roswell Museum, or on Dennis Balthaser in person. This can't be ruled out. For a jape it was a convoluted one, requiring the involvement of a number of different people over a period of some months, but this would just make it all the more satisfying to pull off. Contrary to what many people believe, hoaxing doesn't always have an obvious purpose beyond being an end in itself, and the UFO community has traditionally been easy prey to those looking for cheap kicks. And what are the disinformation operations of AFOSI and the other dirty tricksters in the intelligence game if they're not big-budget, carefully stage-managed hoaxes, albeit with specific strategic purposes in mind?

Absurd as it may sound, Balthaser's story has the ring of truth. Compared with what happened to Paul Bennewitz, it's really not that outrageous. Was there really a fragment of metal? We don't know. Nor do we know what the point of the operation was. Given the huge popularity of the *X Files* at the time (1997) the fact that the alleged AFOSI agents were a male–female partnership was a nice touch, as was the conspiratorial hint that their boss had recently been assassinated by the CIA. If it was taking inter-service rivalries a little too far for real life, it was exactly the sort of thing that might happen on a television series about FBI agents and extraterrestrials.

What also rings true is the AFOSI agents' choice of meeting place. Espionage 101 advises meeting your contact in public, somewhere crowded and out in the open where there's no opportunity to plant listening devices or other electronic surveillance equipment. And Denny's seems to be a popular choice with UFO insiders. It's where Rick Doty chose to meet Greg Bishop for his interview, it may even be where Doty and Falcon first met Bill Moore back in 1980. As to the statements by Ed, the

AFOSI agent, regarding ETs and UFOs, well, at this point we've heard it all before, though the tracking beacon was a new development and one that doesn't seem to have caught on.

So why would the US Air Force want to promote the Roswell story? We can hazard a guess. By the late 1990s Roswell had become the central pillar of the American UFO mythology: if Roswell fell, then the entire marquee beneath which the Air Force and everyone else carry out their UFO-themed operations might fall, too.

But what AFOSI says to UFO enthusiasts is not necessarily what it says to the rest of the world. Three years previously the Air Force had published its 1,000-page report into the Roswell Incident, written by a former Air Force Special Programs and AFOSI Colonel, Richard Weaver – Rick Doty's former boss. In it Weaver had presented the top-secret Mogul balloon theory as the final word on the matter. As Weaver himself predicted in his accompanying statement, the UFO lobby rejected the report out of hand, leading to a second report being prepared, *The Roswell Report: Case Closed*, in which the aliens described by witnesses were explained away as confabulated memories of crash-test dummies. This second report was released in June 1997, within a couple of weeks of Balthaser's meeting with the alleged AFOSI agents.

One is left with the impression that the US Air Force operates a two-channel system. One channel is for the general public, those for whom inoculation against the alien meme might work. They're encouraged not to take the Roswell UFO story too seriously, though visiting the city and the museum is probably regarded as harmless fun. The other channel is for those already infected with the UFO virus and these people, the incurables, are sporadically encouraged in their beliefs, perhaps to keep the story alive until it's next needed for an intelligence operation. Some of their names we now know – Silas Newton, Olavos Fontes, Bob Emenegger, Paul Bennewitz, William Moore, Linda Moulton Howe, Bill Ryan, Dennis Balthaser – but we can be sure that there are many more out there.

THE SECRET LIFE OF WALTER BOSLEY

A heavy-set, jovial and gregarious man in his forties, Walter Bosley is an instantly likeable character, one you might expect to find running a comic or gaming shop. With his near-waist-length hair and general uber-geek vibe he'd be a model dungeon master in your perfect *Dungeons and Dragons* scenario, which is appropriate for a man who spends part of his leisure time scouring California's deserts for entrances to the Hollow Earth. Walt, as he likes to be known, also writes and publishes old-fashioned pulp fantasy and science fiction under a pseudonym and has written non-fiction about the Masonic symbolism at California's Disneyland. When we met Walt, in a town absorbed by the endless sprawl of Los Angeles, he had been engaged with somewhat more mundane work, running security checks on would-be contractors at Edwards Air Force Base, a position of trust afforded him by his decade's service for the FBI and AFOSI.

A powerful out-of-body experience as a child left Walt convinced that there was more to this world than was immediately apparent. It was a feeling that lingered with him throughout his adolescence, leading him to read voraciously about ghosts, UFOs and other denizens of reality's fringe. A few years spent living in West Virginia during the outbreak of high strangeness documented in John Keel's *The Mothman Prophecies* only encouraged him further. During these formative years Walt's mentor was a relative who served an entire career with the FBI. It was he who told the inquisitive youngster that if he wanted to know the truth about UFOs then he should consider joining the Bureau, which he did in 1988 after completing a degree in journalism. Walt spent five years with the FBI, learning Russian, training as a surveillance specialist and conducting physical, then telephone surveillance operations.

In 1993 Walt shifted gears and entered AFOSI as a special agent, working from Wright Patterson AFB; he had hoped to learn the secrets of Hangar 18, and sounded disappointed when he told us that if there had been any ET technology at Wright Patterson, it wasn't there now. Walt was chief of Wright Patterson's Counter-espionage branch, where he monitored another type of UFO – that is, Unidentified Foreign Operatives. According to his resumé, Walt 'managed, designed and conducted double-agent operations in international venues, from mission objectives directed on a national level, in cooperation with the FBI and CIA'.

There's a lot to be gleaned from even this brief job description. It immediately dispels the idea that Rick Doty was a rogue operator: AFOSI counter-espionage projects, it tells us, are directed from above and operate on a national level. I wondered whether Kirtland AFOSI's operation against Paul Bennewitz and the UFO community ever came up during Walt's own training. It did, Walt remembered, though more as an example of how *not* to do things. We can also see from Walt's resumé that AFOSI were running their operations in collaboration with the CIA and the FBI, bolstering Doty's claims that the NSA were involved in the Bennewitz operation. Walt himself made the smooth transition from counter-intelligence in the FBI to doing similar work for AFOSI, suggesting that there's a fluid overlap in the work that all these agencies do.

Walt admits that part of his motivation for moving from the FBI to AFOSI at Wright Patterson was to find out what was really going on in the UFO arena; and in a way, he did. At both the FBI and AFOSI, Walt learned that he was not alone in his fascination with the paranormal. Like any large organization, these agencies are whirling microcosms of our wider society, albeit better trained and better equipped, if not always better paid. Like any society, they have their fair share of religious advocates of many stripes, and UFO and mystery enthusiasts are prominent among them. Unlike Rick Doty, Walt didn't get a glimpse at the Red and Yellow Books – the repositories of ET knowledge locked away in the government's vaults. The way he acquired information about the UFO situation was more subtle: as would befit employees of a clandestine information-gathering organization, once people knew that you shared an interest in the strange, *they* found *you*. Walt told us that 'it would be a case of somebody sidling up next to you in the canteen and saying, "Hey, so you know where the UFOs come from?" They come from the Hollow Earth, or Antarctica, or wherever . . .'

Most of the stories Walt heard were friend-of-a-friend playground whispers, but when you're in the intelligence business, you're in one of the most exciting playgrounds in the world, and your friends are the people who are supposed to know what's really going on out there. Even if their information is second- or third-hand, inside knowledge from people you might one day rely on to save your bacon has a much great currency than random scatter pulled off the Internet.

It's this innate sense of trust within the informal, internal networks of military and intelligence personnel that has allowed the UFO myth to survive and flourish there unhindered for so long. It's why Apollo astronaut Ed Mitchell can publicly state that he knows UFOs and extraterrestrials have visited the planet, even though he has never seen them himself – to which everyone will listen. Walt and Ed Mitchell trust the people who told them that the ETs are here and, as insiders, we believe them.

At Wright Patterson and AFOSI, the UFO stories intensified, but by then Walt was also helping to spread them in his role as a special agent. UFO-themed operations were by no means all that AFOSI did, however; they were just one of a number of projects that Walt would be keeping tabs on at any time, but they were a recognized category within the wider counter-intelligence programme and one for which Walt, with his predisposition towards the material, had a natural propensity.

What exactly he was doing Walt couldn't tell us, only that it provided cover for a particular highly classified aircraft if it was spotted by anyone who wasn't supposed to know that it existed. And that meant *everyone*, including Walt. Even he didn't know what the damned aircraft was, only that he was watching its back and encouraging people to believe that it was not of this Earth.

After almost two years of this, Walter cracked – he *had* to know what he was hiding. He asked one of his Air Force liaisons if he could see the aircraft, arguing that it would help him do his job better. A few phone calls were made, Walt's immaculate service record was checked out and a few days later he got a phone call. He could see the aircraft. A time and a place were given to him: he was to rendezvous with another operative in the middle of the night, in the middle of nowhere.

On the given date, Walt drove out into the night. An Air Force jeep was waiting for him in the dark at the agreed location. He pulled up alongside it, got out of his car and greeted the other vehicle's driver.

'So, when's the plane due?' asked Walt.

'It's already here,' was the reply. 'Look up.'

Walt looked up but saw nothing but the starlit sky above him. The night was silent apart from their conversation and the dull whoosh of cars passing in the distance.

'I don't see anything,' said Walter, puzzled. 'Where should I be looking?'

'Just look up,' said the driver.

Walt looked up. There was nothing to see. He realized that he'd been pranked. The message was obvious – you don't see what you're not supposed to see. It was time to go back home.

'Oh, wait,' said the man. 'I forgot to give you these.'

He handed Walt a pair of goggles. Walt put them on.

'Now look up.'

Walt looked up.

'Holy fucking shit!'

As he told his story I could see that Walt still felt the surge of adrenaline that had rushed through him, almost knocking him to the ground.

There it was. Hanging in the air so low that he could almost touch it. Totally silent. Totally invisible. Totally insane.

Walt recalled the only thought that ran through his mind: 'That's the coolest damned thing I have ever seen . . . '

Several years later, Walt was still buzzing with excitement as he recalled the scene. A scene that, naturally, he couldn't tell us anything about.

'So what did it look like?' I asked.

'Oh come on, you know I can't tell you – and I'm not going to.'

'Was it triangular? Spherical? Elliptical? Rectangular? A boomerang?'

'Look, I'm not going to tell you anything; other than that it was amazing. I'm not even sure that I could describe it if I wanted to. And I don't think that anyone could guess what it was either. Just believe me when I tell you that we've got some crazy shit flying up there. And I think some of it may not be ours . . .'

'Hang on,' I stopped him. 'What do you mean, "not ours"?'

'I mean the technology isn't ours. It's not human . . . Not our kind of human anyway.'

I pressed Walt for more information, feeling the vertiginous rush of approaching strangeness.

'Um, how weird can I get?' Walt asked. It seemed like a rhetorical question.

'As weird as you need to be,' I replied. And Walt got weird.

'Well, you see . . . and this is all stuff that I heard from people I was working with in the Air Force . . . they . . . the others – the people with the UFOs – they're not exactly alien . . . well, some of them are . . . but the ones we deal with . . . they're, well . . . they're *us* . . . they're humans . . . but us from another star system, Sirius . . . they're time-travellers . . . and they seeded life on this planet . . .'

OK, I'd heard this before and could cope with it. The time-travel scenario, once you've made the conceptual leap to a world in which time-travel exists, is a reasonable one. It would account for the reports of human-like occupants seen on board the craft. It would also explain why, with their immensely superior technology, they haven't yet wiped us off the face of the planet – because we are their grandparents, or their great-grandparents, or their great-great-

great-grandparents, and without us they would cease to exist. It also explains why the government would want to keep the truth about UFOs a secret – imagine the pressure placed on our rulers if they knew that they could change history – or that their enemies could.

I was just getting comfortable with Walt's weirdness. It's actually quite logical in its own way. These were ideas I could live with.

'And it's got something to do with the Nazis.'

John sighed behind me. I realized I was looking at Walt with a quizzical expression. I could hear air rushing into my open mouth.

'I told you it got weird,' he said, sheepishly. 'To be honest I don't know exactly how that connection works, but the Nazis are . . . er, were . . . involved somehow . . . and, I think they live underground . . .'

'OK,' John interrupted. 'That's great Walt. I think we can wrap up the interview now.'

'Did I get too weird?' Walt asked.

'No, but I think we got enough from you and . . .'

'I got too weird, sorry, I do that sometimes. This stuff *is* weird, you know . . .'

We went to a local diner – not a Denny's – for dinner. Walt told us of his doubts about the official story of the 11 September attacks. 'The more you know about how the government works,' he told us, 'the more you find out, the crazier this shit gets.'

And he's right. We'd seen it with Rick Doty and now we'd seen it with Walt. We'd heard it in the stories told by Bob Emenegger and in the tales of other alleged whistleblowers. The closer you are to the source, the stranger the stories get, the more infected you are and the more infectious you become to others. 'UFOs are a mental illness' John Keel once said. Keel was right: UFOs are pathological, and they are highly contagious.

Perhaps exposure to this material, even if you know that you are making up some of it, does affect you on some deep level, destabilizing you, leaving you open to extreme possibilities, unable to separate the signal from the noise. Is this what happened to Walt and to Rick? Could it happen to John and me?

In *The CIA and the Cult of Intelligence* Victor Marchetti and John Marks discuss the problem of 'emotional attachment', which becomes particularly acute for agents working in special operations. They describe a team in the late 1950s training Tibetans loyal to the Dalai Lama for an uprising to reclaim their country from the Chinese, a mission that was fundamentally hopeless and led to many deaths. Several of the CIA trainers later adopted the prayers and beliefs of their charges. Emotional attachment, they note, is particularly prevalent in special operations, whose officers 'often have a deep psychological need to belong and believe. This, coupled with the dangers and hardships they willingly endure, tends to drive them to support extreme causes and seek unattainable goals.'[1]

Is this how it happens? Is there something so deeply appealing, so deeply *right* about the UFO, about the idea of saviours from Outer Space, of technological angels, of our future time-travelling selves, that it ultimately infects everybody that it comes into contact with? Do we need to believe that someone else out there can save us, or at least give us hope that we, as a species, as a planet, can survive the perpetual chaos of life on earth?

Inherent in the UFO mythos is a conflict of common sense and paradoxical nonsense – a cognitive dissonance – that is so heady, so potent, that speaks to us on such a deep level, that it opens a route directly to the soul. Is this what the governments want to protect their people from? Why they must constantly say 'NO', even if they don't really have an answer? Because the material is nothing short of toxic?

I think so, and when carriers such as Rick Doty and Walter Bosley get into the corridors of power, as they sometimes do, then there's every possibility that their infection, this extraterrestrial virus, might spread. And from there it wouldn't take much for the contagion to get dangerously out of hand.

SEVENTEEN

DOWN TO EARTH

'Have you tested this theory?'
'I find it works well enough to get me from one planet to another'
Professor Barnhardt to Klaatu in The Day the Earth Stood Still *(1951)*

Whatever strange game the US Air Force had been playing with the UFO community for the past six decades, talking to Walter Bosley and Dennis Balthaser persuaded John and I that it was not yet over. My hunch was that it is never supposed to end: the 'truth' about UFOs will always be just around the corner but, as in a bad dream, that corner never seems to get any closer no matter how fast you run. We were still wondering how Serpo fitted into the jigsaw. We'd seen the catalytic effect that it had had on the UFO scene, but we still had no idea where it had come from or, indeed, what its purpose was – if it had one beyond extending the grand ufological narrative a step further.

We didn't have much longer to consider it, however. Things were not going well in the Serpo camp, and as the long-promised photographs of Eben ball games still hadn't shown up, the audience was getting restless. There also seemed to have been a breakdown in communications between Request Anonymous and his messengers.

To make matters worse, it was starting to look as if the alleged sources for the Serpo material – Anonymous, Paul McGovern and Gene Lakes – were all multiple personae of one man or, at least, one computer. The discovery was made by Steven Broadbent, a British computer network specialist who had been drawn into Serpo's embrace and had set up a website, *Reality Uncovered*, to examine the story as it unfolded. After engaging in some online cloak-and-dagger work, Broadbent found that the original Anonymous emails distributed to Victor Martinez's email list in November 2005, and the corroborating emails sent by Rick Doty, Paul McGovern and at least two other alleged Defence Intelligence Agency 'insiders', all shared the same consumer broadband IP address – which identifies the local computer network that a computer is using – based in Albuquerque, New Mexico. This fingered Rick, the only member of the group known actually to exist outside of cyberspace, as the probable sender of the original emails, if not the source of the Serpo material itself.

It seemed incredible; could that really be Rick, chuckling backstage on planet Serpo? And, if it was, who, if anyone, was behind him? Broadbent's findings instigated a flurry of forum postings and email exchanges that continued for several months, all seeking to identify the minds behind the Serpo saga. Was it just Rick? Was Victor Martinez involved? What about the other members of Bill Moore's Aviary? Was it a branch of the US military, the intelligence services or the government? Was it Scientology? Could it have been the publishing company behind Rick and Bob Collins's *Exempt from Disclosure*, or one of the UFO and conspiracy websites that had sprung up in Serpo's wake, such as *Open Minds* or *Above Top Secret*?

It was a mystery to make Sherlock Holmes unplug his modem in despair. The cast of characters was lengthy, the motives given to each by their accusers – usually someone from one of the other groups under suspicion – were rarely convincing and, unless somebody stepped forward with a full confession, the case appeared to have collapsed under the sheer weight of information, misinformation and disinformation loaded upon it.

Broadbent's findings appeared to show that Rick Doty had sent Victor Martinez the original 'Anonymous' postings that had set the Serpo machine in motion; Rick had then followed them up with corroborative emails from himself and the other fake Pentagon UFO insiders. Alternatively, somebody had actively manipulated these emails to make it look as if Rick had sent them – something well within the bounds of possibility, especially if there was an intelligence component to the Serpo saga.

It was difficult for us to believe that Rick could have generated the voluminous Serpo material alone, though he may have had a hand in it. If Rick was responsible, and he was denying it to the bitter end, why might he have done it? Could Rick have started out taking on what looked like straightforward freelance intelligence work, then found himself dropped in the shit as a patsy when things got messy? Or was it a mess of his own making? Might what he knows about the origins of the Serpo material make it impossible for him to tell the whole story, even to save his already shredded reputation? Or was it his reputation as a UFO trickster that got him selected for the task in the first place? A great black hole was threatening to engulf the entire Serpo saga; would it take Rick with it?

THE CONDUIT

Serpo's death spasms were ongoing while John and I were in Los Angeles, so we sought out Victor Martinez, the original distributor of the Serpo messages, thinking that meeting him might shed some light on the affair. As we should by now have expected, our encounter only raised more questions.

A stocky, barrel-chested man in his early fifties, sporting large dark glasses and a neat moustache, Victor Martinez speaks quickly and at great length, waving his arms around dramatically to emphasize a point. Victor lives alone in a family home in Pasadena and occasionally works as a supply teacher. He suffers from chronic ill health – he was wearing a hospital wrist band when we met him – and is a stickler for spelling, grammar and punctuation. Several times as we talked in Greg Bishop's garage, Victor noticed loose nails on the floor, picked them up and pocketed them anxiously. He's also something of a technophobe, which is odd for a man who maintains such a large and influential email list. At the time of the Serpo affair Victor didn't even have a computer at home, running his operation from a basic web-TV set-up.

What's not clear about Victor until you meet him is that he's an ardent, lifelong believer in ET visitations to our planet; he also believes that the ETs seeded life here and that the government is covering up these facts. He has been actively involved in Los Angeles area UFO groups since at least the early 1980s, during which time he knew Bill Moore, a connection that might perhaps have raised a few eyebrows if anybody had known about it.

Victor loved his new-found status as a UFO insider – and who wouldn't? Being able to pick up the phone to Rick Doty or Robert Collins to discuss the latest UFO developments would be like speed dialling Michelle Obama for a blast of White House gossip. And, despite everything that had happened since he first allowed his list to be Serpo's birthing pool, he still trusted Rick. Victor referred to Rick as 'a great friend', and remained sure that at the root of the Serpo story was a truth that needed to be told. In other words Victor was a perfect mark for an unscrupulous player seeking to inject new DNA into the ufological bloodstream.

Much as we liked Rick, it was getting more difficult for us to defend his role in what was fast becoming known as the Serpo hoax. Fresh heat was being directed his way, and now we were being caught in its path. Intelligence players on the UFO scene were accusing Rick of acting illegally by taking us on to Kirtland AFB. In reality there was nothing illegal about the act, unless I was a foreign agent – and somebody was still doing their best to paint me as one, and so cause trouble for Rick. At times it seemed that John and I were the only people trying to stand up for Rick. But as it turned out, we weren't.

THE MAN IN THE WHITE COAT

When you get a summons from a former CIA scientist, especially one who's had a foot in the UFO arena for some thirty years, you tend to answer it. That was how John and I found ourselves on one of the most hair-raising rides of our lives, speeding across the post-apocalyptic wasteland of downtown Detroit in a yellow cab driven by what I can only accurately describe as a homeless man. He may not have been entirely homeless – he probably slept in his cab – but his clothes were covered with the greasy sheen of human oils, his toes were poking out of the ends of his filthy shoes and, well, he stank. Our cabbie was driving at sixty miles an hour down the freeway while using both hands to huff something out of a brown paper bag. It may have helped to focus his attention, but the situation became alarming enough for me to ask him, gently, please to keep at least one hand on the wheel.

When we eventually reached our destination, the receptionist couldn't find the man we wanted to meet in the staff directory, but that was OK because, unnervingly, he was already in the foyer, dressed in a white doctor's coat and greeting us with a firm, hearty handshake and a welcoming grin.

Kit Green is a formidable, immediately likeable character and everything you would want from a former CIA employee. Sharp as a quill and as carefully composed as a legal letter, he is clearly a minimal-bullshit kind of guy. As a medical doctor specializing in neurophysiology and its applications to national security, you would hope so. He's also fascinated by UFOs, an interest that goes back some three decades to his time as the CIA's 'keeper of the weird' at the Office for Science and Technology. Really he was a senior science analyst, but a small part of his work involved intelligence to do with remote viewing (RV), UFOs and other hot paranormal topics of the 1970s.

And Kit still seems to be very much in the loop when it comes to the business of UFOs and the US government. His take on the problem is refreshingly complex and nuanced, recognizing that while UFOs can end up being a form of mental illness for some people, there is something genuinely strange and worthy of investigation lurking at the heart of the mythology.

Kit, John and I spoke for some hours, initially about the Serpo material, which, to our surprise, Kit took more seriously than we had expected. 'There are certain facts in there,' he told us, 'certain references, that prevent me from being able to reject the material out of hand, even if the story that it's telling is patently not true.' The Serpo material, or at least some of it, Kit suggested, might have served a purpose to someone, somewhere, perhaps conveying information in heavily codified form. One of the ways you can assess the value of information is to watch who is drawn to it, and Serpo had caught the attention of some senior players in the defence intelligence field – perhaps some of the most senior players. And that was enough to keep Kit interested but, like the rest of us, he still didn't have any answers.

Kit is also a close and long-standing friend of Rick Doty, who he talked about with unguarded warmth and respect, though he was forced to admit that sometimes Rick's actions could be both puzzling and frustrating.

Given that I now had a direct audience with a man who, at least, ought to know what's going on behind the scenes in the UFO arena, I decided to throw a few straight questions his way. And I got some admirably straight answers.

I asked Kit about the silver spheres from Yosemite – did he have any ideas what they might be?

'We have reconnaissance craft that fit that description,' he said matter-of-factly, though he wouldn't be pressed for further information.

We talked about the use of quiet helicopters 'disguised' as UFOs to probe security and nuclear installations.

'I'm pretty sure that it happened,' said Kit, following up by hinting that he may have met a man who claimed to have flown similar missions.

That was two strikes against the extraterrestrials. But just as we were entering our conspiratorial comfort zone, Kit soon hauled aliens into the equation. With his doctor's knowledge of biology, his analyst's head for information, his CIA officer's understanding of security matters and his UFO buff's capacity for high strangeness, Kit is sometimes passed curious material for assessment.

Back before the hoax 'alien autopsy' film made by Londoner Ray Santilli made global headlines in 1995, Kit had heard tell of reports detailing autopsies performed on beings that weren't quite human. They're just one part of the body of material that makes it impossible for him to reject the possibility of ET visitation out of hand, and why at a Denny's restaurant back in 1986 he, along with physicist Hal Puthoff and computer scientist and ufologist Jacques Vallée, distilled what they knew about the subject into what has become known as the 'core story'.

Simply put, the core story, according to Kit, is this: 'The ETs came here, maybe once, maybe a few times. Either through accident or design, the US government acquired one of their craft. The only problem was that the physics that powered the craft were so advanced that for decades we humans have struggled to understand it or to replicate it.'

It may not have been much, but, hearing this information from a man of Kit's calibre and background, the effect was akin to a small nuclear device going off in your head. I asked him to repeat his statement, and he did – the aliens have been here. Kit casually mentioned the nickname the 'three bears', given to three of the ET craft allegedly in US possession. In a flash I remembered a puzzling email I'd received from Rick Doty. It contained a text titled 'The Three Bears' that described, in broad terms, an Air Force mission to extract American spies from the Soviet Union. Was the story really about something else? My mind lurched uneasily as I felt pieces of a larger puzzle slide into place.

This man was serious. And, as he sees it, so is the situation. We are not alone and probably never have been. They may even be here now. The problem is, what are we supposed to do about it?

Kit is acutely aware of the problem and he may have the solution. It emerged in a lengthy monologue, one I suspected he had given before.

'In a country that has a large, educated population there is a large subset of individuals who suffer from what's called paraphrenia. Paraphrenia is a form of mental illness that doesn't interfere with your everyday life. It means that you can have a delusion and not be crazy, a delusion that you can confine and control. Many of us have one corner of the mind that is delusional – I bet you that I do.

'It might, for example, be religious – I'm an Episcopalian, though as such I'm protected from diagnosis, as are all the UFO buffs, because a large social structure of shared beliefs, like a religion, cannot be a delusion. So all those people who believe that they are being beamed at by the government can no longer be diagnosed as crazy – there are just too many of them.

'But, if there is a condition that is threatening to the social structure – like the idea that the aliens are here and they are taking our babies, or that God hates people of a certain creed or colour – and if people who believe that kind of delusion band together, they can end up encouraging each other to get a lot sicker. And, once in a while, they end up committing mass suicide, or they strap on belts and make themselves human bombs. So we have to know how to deal with these people and how to prevent them from being dangerous to others.

'This applies to the UFO problem. If something really strange in the area of UFOs is true, then what do we do about conveying that information to the public? First we consider what may be the basic facts: maybe there are civilized lifeforms elsewhere in the universe; maybe they visited us in their spaceships a couple of times and then went back home; perhaps they left a vehicle or some technology behind and we've spent a lot of time and money trying to figure out how to use it. And there may be people in the government who believe that this did happen, and believe that the information needs to be public knowledge, because perhaps someone outside of the government will be able to make sense of their technology. But then there's another group of people in power who say, "No, it will make them sick to know all this, we can't let the story out, it's too dangerous."'

John and I glanced at each other. My mouth was dry. I felt the temperature drop. Or was it rising? I wasn't sure. Things were getting strange again. Did Kit just tell us that he knew that these things had happened? Was that a hypothetical scenario he had just presented us with, or one that he believed to be real? Kit continued.

'So, what do we do? There are studies on both sides of the problem. Some show that people will go crazy and jump off bridges when they're presented with this information. Others, however, say that if you don't want them to go crazy, what you do is you systematically desensitize their fears.

'If you are a psychiatrist with a patient you can do that in a very methodical way. If you are a sociologist working with a group of students at a university you can also do this in a very structured and experimental way. But if you are a government working with a population it's a lot more complicated. Sure, there are those who are just going to shrug and say, "I always knew the aliens were real, it's no big deal." But you also know that some of them are nuttier than a fruitcake and could cause a lot of trouble. So we have to ask ourselves how we can tell people what they deserve to know and, maybe, what they *need* to know?

The way to do it is to construct a framework whereby they can parse out the things they've heard that are not true, and you whittle it down to a manageable story. A story like this: "There were three spaceships that came here over thirty years, and we've got one of them. We can't figure out how it works, we've crashed it because there's a lot of physics that we've still got to learn. We do have something that's like a magnetohydrodynamic toroid, and it really did get a craft off the ground, but it smelled bad and it killed a couple of pilots. And we're really sorry about that, but we did it because we've got this machine that came from another planet, and we need to know how it works."'

Oh god, he just did it again. I tried to slow my breathing to prevent the giddiness from becoming a full-on panic attack.

Kit carried on, oblivious to my inner struggle. I was glad not to be inside one of his MRI machines.

'How do you tell people that story? If it's true?' he added, almost parenthetically.

'If you were to give them the core story right off the bat, they'd get sick, so you do it slowly over ten or twenty years. You put out a bunch of movies, a bunch of books, a bunch of stories, a bunch of Internet memes about reptilian aliens eating our children, about all the crazy stuff that we've seen recently in Serpo. Then one day you say, "Hey, all that stuff is nonsense, relax, it's not that bad, you don't have to worry, the reality is this" – and then you give them the real story.

'Now that is a tried and tested desensitization model. And to the extent that there is a core story that is true, and to the extent that the people who want to get the core story out want to do it in a way that won't hurt people, that's what they do – they come up with a bunch of nonsense, and then they get rid of it, gradually making people more comfortable so that when they are finally presented with the truth, it's not as frightening as they had originally feared. "I can believe that," they'll say. "I didn't believe the crazy reptile stuff, I didn't believe that they were abducting babies from wombs on beams of light. But I *can* believe the part about the spaceship and the back-engineered technology."

'And that's what I think the meaning, the rationale, behind things like the Bob Emenegger film is. It's to help people not get sick later, to calm people down when they find out the truth.'

I felt as if I'd been hypnotized. John was looking very serious. Beads of sweat were appearing on his forehead. We'd heard a lot of very strange things on this trip. They were easy enough to categorize, to file away in the Strange Things box and pull out to show people when they wanted to hear something really weird.

But this was different. Kit was different. Here was somebody I would happily depend on to save my life. As a doctor, Kit has probably had to do that a few times. Here was a very sane, very intelligent man, a man who had been closer to the secret machinations of government than anyone we were ever likely to meet. And he appeared to be telling us that the aliens are real. And that they've been here. And . . .

I swallowed down the creeping sense of panic, of the solid ground on which I'd built my worldview vanishing beneath my feet. In the many years I'd spent absorbed in the UFO subject I'd vacillated between wide-eyed optimism and hardened scepticism, settling on something awkwardly in between. While I'd never completely denied that tiny seed of cosmic hope, I'd also kept it sterile and unnourished. And now I could feel it stirring, ready to unfurl its great leaves, block the warm, clear light of certainty and bathe me in the cool shadow of doubt.

If Kit could believe, then surely I could, too.

John and I had been with Kit for most of a day now, and it was time for us to leave. This was probably a good thing: I wasn't sure how much more of his quietly sensible shock and awe we could take. We thanked Kit for his time and he called us a taxi.

Whilst he wasn't huffing from a brown paper bag, our new driver was still leeringly insane. And from the dazed, startled looks on my and John's faces, you'd be excused for thinking that we were, too.

A DIFFICULT YEAR

A few days later we were back in Albuquerque. To our surprise we heard from Rick Doty, who was keen to catch up with us. We arranged to meet in a Denny's at the shopping precinct in his home town, about an hour's drive away. Denny's again: I wondered whether all Denny's restaurants are wired up by the National Security Agency, making instant access to your conversations a simple matter for intelligence operators.

John and I waited outside the restaurant for Rick. An Army helicopter was installed near the supermarket on the other side of the parking lot. Kids, some of them future American warriors, clambered in and out clutching soft drinks and ice creams.

A huge silver jeep the size of a small tank pulled up alongside us, Rick grinned from behind the wheel.

'Hi guys! You're not filming me, are you?!'

We assured Rick that we'd sent the camera crew packing for the day. He seemed genuinely pleased to see us, and we were genuinely pleased to see him. He was definitely off duty, in a baseball cap, t-shirt, shorts and trainers. One of his eyes was puffed up; it looked infected.

'Yeah, I have a sinus infection; it's a little uncomfortable. It's been a rough year. There have been a lot of health problems in my family.'

Rick and John ate salad while I tucked in to a plate of prime rib. Rick seemed relaxed.

After some initial chit-chat we got talking UFOs. Rick talked about Project Silverbug, about flying saucer test flights over Canada back in the 1950s and 1960s, about ongoing American saucer testing and funding connections to central European royalty. He talked about real flying saucers landing in Coyote Canyon at Kirtland in the 1970s. Was this what Bennewitz saw? Maybe.

Suddenly, in typical Rick style, we were off the deep end. He told us that one of his jobs at OSI was providing security for a colonel who was the key liaison with EBE 1 (he pronounced it 'eebah'), the extraterrestrial who had stayed behind. The colonel was now dead, but his daughter had found some black-and-white photographs of her father with EBE 1. If the photos were found to be authentic, Rick said, then they would be published.

I gave Rick my best incredulous look. He remained deadpan. I wondered if he was rehearsing a new story with us, a new routine.

Where did he get this material from? Does he prepare it? Does someone feed it to him? Does he spontaneously generate outrageous UFO material? Was Rick an unrecognized genius of postmodern fiction? Had he penned the Serpo documents in order deliberately to blur the line between consensus reality and the social imagination; to expose all history as myth? He should be a science-fiction writer, I thought, not a policeman.

Rick told us that back in the 1970s, when Hal Puthoff and Russell Targ were running their Remote Viewing research at the Stanford Research Institute, they had asked their psychics to look at another planet. Their description, Rick told us, included adobe-style huts and alien beings. Just like Serpo. He also told us that the business of the dubious email addresses was a frame-up. John and I agreed that this was possible, but that he had to understand that his situation didn't look good to the rest of the world.

We talked like this for a while, Rick dropping crumbs and clues, some of which I can't repeat here. Knowing how Rick liked to talk tech, I asked him about the holographic technology described by Bob Emenegger – did it have anything to do with the UFO puzzle? Rick became reticent and changed the subject.

John and I told him how since starting on the project we'd got a better sense of how the UFO rumour mill works, how pieces of the puzzle will suddenly slot into place months or years after you first heard them. Rick smiled approvingly, like a mentor pleased with his apprentices. He told us that nobody he knew had the full picture, and that's the way it's supposed to be. That's the way secrecy works: everything is compartmentalized, then mixed with such a knotty tangle of disinformation that no one person will ever find their way to the heart of the labyrinth.

I suggested that the truth may now be so deeply buried that nobody, anywhere has the answer. Rick disagreed; he felt sure that someone has the blueprints – he just didn't know who. It certainly wasn't him: 'I'm just a pea in a pod,' he said.

A couple of hours passed, then it was time for us all to move on. We said friendly goodbyes and invited Rick to visit the UK some time – he told us that he'd always wanted to visit Loch Ness and hoped to do so one day.

As he drove away, his enormous silver machine vanishing in clouds of desert dust, I realized that I was very fond of Rick. He may be a deceiver, he may even be dangerous in some way, but above all else he was an alien to us, an enigma wrapped in a space blanket, tucked inside a flying saucer.

But is it one of ours? Or one of theirs?

HOW TO BELIEVE WEIRD THINGS

The following night, back in Albuquerque, Greg Bishop and I retired to a bar for a beer or three to talk UFOs and space music, surrounded by the reassuring, earthly clink-clank of bottles and billiard balls.

Greg has been involved in the American UFO scene for a good two decades, and when it comes to the human side of the phenomenon, he has seen it all. Although he's never had a UFO encounter of his own, he's seen a lot of people get hit hard by UFOria. He remembered William Moore, lying prone on his back in a state of near panic as he told Greg that the aliens were here on Earth. He recalled Jamie Shandera, Moore's colleague in the MJ-12 and Aviary investigations, transformed suddenly from cynical disbeliever to terrified convert. When pressed for a reason, Shandera would only say 'they showed me something'.

Greg himself was overcome with paranoia during the mid-1990s. At the time he had begun receiving UFO information from a man who presented himself as a Naval Intelligence source. Greg became convinced that he was being monitored and bugged, that government agents were in a van parked outside his home, extracting data from his computer. He began hiding the audio cassettes of his interviews with the Naval man inside a biscuit tin. Fear ruled his life until, one day, he decided that enough was enough and refused to feed the fear any further. As Greg will tell you, he's been completely sane ever since.

With a *soupçon* of satisfaction in his voice, Greg told me that he'd noted a change in John and me since we set out on our mission. 'When you started out neither of you were really taking this material seriously,' he said. 'You laughed at the stories you were hearing, you made light of what was going on around you. Now I sense something different in you both. You're still laughing, but *with* the phenomenon, not *at* it. The lightness has gone. Your certainties have been challenged. You're having to consider carefully what it is that you really believe. Well done! You're entering what occultists call the abyss, you are stepping out into Chapel Perilous. It feels good, doesn't it?!'

I thought back on whom we'd spoken to. Rick, Walter, Kit. They were the Mirage Men, our own Three Bears; men at various levels of inside knowledge, trained in the techniques of deception, skilled in the art of secrecy, telling us things that ranged on a sliding scale from the outlandish to the bizarre. And yet they appeared to believe them.

Were they deceiving us when they matter-of-factly told us that the US government is flying extraterrestrial technology? Had they themselves been deceived by someone else, or were they deceiving themselves? Or do we just have to accept the fact, however uncomfortable, that somewhere out there in the world's great wildernesses, something deeply strange, something that is not of us, is gliding at incredible speeds through our atmosphere?

Whom can we believe? Rick and his Yellow Book, the holographic repository of all alien knowledge? Walt and his time-travelling, subterranean humanoids? Kit and the core story? What can you believe and still retain your sanity? Greg and I agreed that the only way to deal safely with these questions is to make peace with the fact that you will probably never know the answers.

We sank a couple more beers and traded tales of high strangeness. Greg told me how a former government remote viewer (RV) had once psychically attacked him over dinner, projecting images of violent assault into his mind. I told him that I once saw, momentarily, a colossal flying saucer hovering over the city of Austin, Texas, as I was driving on a nearby hill. It looked like something from the film *Independence Day* (released a year later) and must have been thousands of feet across.

We were toasting our own unshakeable sanity when John loped into the bar, smiling strangely and clutching a bottle of beer.

'Hey guys!' He sat down. He seemed awkward, unsure of himself. He paused as if contemplating his next move.

'Listen guys, I need to talk to you. Actually, I think I need your help . . .'

'Sure, what's up? Are you OK?'

'I think I'm OK, but . . . something has been bothering me since we saw Kit. I'm . . . er . . . well I'm starting to believe this stuff; you know, the aliens, the UFOs, the government cover-up. All of it. Until we met Kit I thought I knew what was going on. When it comes to crop circles I *do* know what's going on. But now, I just don't know. . . . If Kit believes that they have an alien craft, that it's all real . . . Shit! I really didn't expect this to happen!'

We told John not to worry. Belief, like love, hate and fear, is one of the most powerful psychoactive agents in the human emotional armoury, and this was particularly strong material we were dealing with.

It was straightforward for us to listen to Walter or Rick talk about their wilder notions, because they were easy for us to reject out of hand as crazy talk. Rick and Walter, for all their experience and knowledge, don't carry the same personal and institutional authority as Kit, a doctor at a large hospital, speaking to us wearing a white coat. Remember Stanley Milgram's 1960s experiments into obedience and authority? Just dressing somebody in a white coat immediately imbues them with psychological power over others. The power to be obeyed, to be believed. Whether intentionally or not, Kit conveys that same respect – through his presence, his articulacy and his knowledge.

I reminded John that Kit told us he was a Christian, an Episcopalian. No less strange a set of beliefs than UFOs and aliens; one that is, if anything, more illogical than the idea of life on other planets. And yet we don't think of Kit as strange, for the precise reason he explained to us: if enough people believe in something, however bizarre, they can't be considered crazy. We might think of them as being infected with a meme or an information virus, but as long as they are functioning within society, if they leave us alone, we will leave them alone.

There's an art to believing weird things. The White Queen in *Alice Through the Looking Glass* believed as many as six impossible things before breakfast, insulating herself from the kind of culture shock that John and I had been through. Some modern schools of occult practice encourage you to adopt contradictory belief systems as a means of understanding how belief works.

Seeking absolute truths will always land you in ontological trouble, no more so than when you are exploring a world as mercurial as that of the UFO believers. Every time you think that you've got a firm grasp on what's going on, it slipslides away from you, leaving you confused and frustrated. The answer, we suggested to John, was to avoid black-and-white, either/or thinking, and to accept that we were exploring a world of gently shifting greys.

Greg told us how Pyrrhonism had been a useful framework for him while exploring these liminal zones. Introduced to him by Marcello Truzzi, a late, revered researcher and philosopher of anomalies, Pyrrhonism was a school of ultimate scepticism from first century B.C. Greece. It taught that nothing can truly be known, that everything must be questioned and that happiness can be found only by avoiding dogmatic thinking and experiencing the world in a state of permanent enquiry. It's a concern that has always bubbled through the history of Western philosophy, surfacing in the agonies of Immanuel Kant and the musings of science satirist Charles Fort before finding new proponents in the extreme relativism of the postmodernists. Pure, honest sceptics, the Pyrrhonists make a clear distinction between a lack of belief in anything and disbelief in something, carefully walking the tightrope between the demon-haunted world and the world devoid of wonder.

To survive in these strange territories, we told John, we must become ontological astronauts, aliens clad in Pyrrhonian armour, surveying the ever-shifting landscape from our silvery airships, borne aloft by imagination and ballasted by doubt.

By the end of the evening the combination of beer and gentle deprogramming seemed to have worked and John appeared calm again. A good thing, too, as we still had one more stop to make before reaching our journey's end.

THE ALAMOGORDO INCIDENT

Sandwiched between the verdant heights of the Lincoln National Forest and the dazzling, gypsum dunes of White Sands National Park, Alamogordo was built as a railroad town in the late nineteenth century, and now services America's vertical expansion into space.

Its motto is 'The Friendliest Place on Earth', but that really depends whose side you're on. This small, unlovely roadside city, the final destination on our flying saucer pilgrimage, is at the epicentre of America's Cold War landscape. Trinity, the site of the world's first atomic bomb detonation, is sixty miles away to the north west, part of the White Sands Missile Testing Range that abuts nearby Holloman Air Force Base, location of the alleged saucer landing documented in Bob Emenegger's film and the Serpo papers. Albuquerque and Los Alamos are about 150 miles north as the saucer flies, while Roswell is about eighty miles or so to the north east. This unearthly, barren landscape is where so much of the early UFO mythology took shape, its alien heartland.

Until 2006 Holloman was the home of the F117-A Stealth Fighter. When I first visited the area in the mid-1990s, those large, ungainly-looking flying arrowheads regularly swooped over my head as I clambered among the alien dunes at White Sands. These days Holloman houses the Stealth's costly and controversial successor, the F-22 Raptor, built, like the F117-A, by Lockheed Martin. In the post-Cold War climate, politicians have argued that the F-22s, intended primarily for air-to-air combat, are already largely obsolete, because no other nation has an aircraft to touch it. This shift in emphasis in America's aerial strategy is clearly visible at Holloman, which also serves as the US Air Force's main training ground for its ever-growing fleet of Unmanned Aerial Vehicles, like the MQ-1 Predator, the MQ-9 Reaper (both 'hunter-killers'), and the RQ-4 Global Hawk surveillance drone, hi-tech descendants of the venerable Lightning Bug.

John and I needed a few shots of the base's main entrance off Highway 70. I wandered across the public parking lot to the guard house and asked permission to film. Permission granted, we set up the camera on its tripod and began to film what we wanted.

I was standing behind the camera, filming passing traffic when something caught my eye, hanging beneath a highway overpass perhaps a hundred feet ahead of me. At first I thought it must be dangling on a fine thread from the roof of the overpass; perhaps, I thought with a shudder, it was an extremely large spider. I immediately realized that this was ridiculous: I could see no thread and the spider would have to have been a foot wide to appear so large at that distance.

I wondered if the thing was sticking up from the ground, poking out above the scrubby reed grass beneath the overpass. But I saw no way in which the object could be supported. The reeds were waving slightly, and the thing wasn't. In fact it was totally immobile. After staring intently in its direction for a few moments, I also realized that it wasn't beneath the underpass at all. It was hovering, completely motionless, in the sky in the far distance.

I stared as hard as I could, but could see no rotor blades, nor could I hear anything but the occasional passing car. It wasn't a helicopter. I turned the camera round, peered through the viewfinder and zoomed in as far as was possible. The object was too far away to see clearly, but it began to take on an oblique form; dark and sinister-looking. I remembered feeling the same way when I saw the Stealth Fighters in the air; they didn't look as if they should fly – too many angles.

Adding to my unease, this thing was also completely motionless. I realized that I'd never seen something so still in the sky. I filmed the object for a couple of minutes and, when nothing happened, stopped recording, deciding to save our batteries and tape. I called John over, taking my eye away from the viewfinder for no more than three seconds as I looked in his direction.

'John, come and look at thi—' I turned back to the camera. The object had vanished.

'It's gone!' I cried incredulously. 'But it can't have done, it was there just seconds ago – how can it have disappeared so fast?!'

We looked all over the clear blue sky, but there was nothing to be seen. Had it zipped away at impossible speed? Had it winked out of sight, perhaps switching on its optical camouflage systems? Had it been watching us filming it?

I was furious for having stopped the recording just moments earlier – had we caught its disappearance on camera then the mystery might have been solved, or perhaps only deepened. Instead the cosmic joker had struck again, presenting us with a genuine UFO.

Here we were on the perimeter of an airbase specializing in advanced unmanned vehicles – exactly where you would expect to see unusual aircraft – and yet we were still stunned by our experience. We were beside ourselves even though, really, nothing much had happened. I had seen a distant black dot hovering in the sky. I had turned around, looked back, and it was no longer there.

Why, then, were we so filled with wonder and glee? For exactly the same reason that the tens of thousands of other witnesses to UFO phenomena have been awestruck by the things that *they* couldn't explain – even if somebody else, crouched inside a cockpit or hunched in front of a computer screen, could.

For all its bathos, our Alamogordo Incident was a perfect reminder of the UFO's ability to shock and amaze, even at its least spectacular. We had finally found what we were looking for.

That night I dreamed of UFOs, three of them. Black seeds against the evening's blue sky. They flipped like dolphins and swung like pendulums, dancing, full of joy, alive and free. They were untouchable, untameable, unknowable. I understood their message. Our journey was over.

EIGHTEEN
WEAPONS OF MASS DECEPTION

1. To exploit enemy superstitions, PSYOP personnel must be certain that:
a. The superstition or belief is real and powerful.
They have the capability of manipulating it to achieve results favorable to the friendly forces

Psyop Policy 36: The Use of Superstitions in Psychological Operations in Vietnam, 10 May 1967

In the months after John and I returned home, the UFOs continued to emerge from their decade-long slumber like drowsy cicadas, falling straight into the arms of an adoring media. Abstract lights in the sky, almost invariably Chinese lanterns, commanded tabloid front pages; astronaut Edik Mitchell announced on the radio[1] that ETs walked among us; numerous cable TV documentaries revisited the ufological canon; an environmentally conscious remake of *The Day the Earth Stood Still* appeared in cinemas and the UFO file at *Fortean Times* magazine was once again thick with press cuttings.

Had Serpo really been the catalyst for this new awakening, or were we just seeing the start of a new cultural cycle, a new generation discovering the power of the UFO for themselves? Either way, the mechanics of UFOria had changed little since my days with the Norfolk UFO Society; each new media account sent more people looking up into the night sky, generating more sightings, which led to new media stories; and so it went on – the circle remained unbroken.

Like one of my youthful NUFOS presentations, our own strange mission had left John and me with more questions than answers, but I felt that we'd got as close to the heart of the mystery as we were ever going to, at least without getting ourselves into serious trouble, either psychologically, or with the authorities. We may not have found any smoking guns, but we'd certainly picked up some spent cartridges and examined a few bullet holes, and the story that emerged from these fragments seemed clear: the US intelligence agencies have been actively involved in the flying saucer arena almost since day one. These people are the Mirage Men, and UFOs are their medium.

Our meetings with Rick Doty, Walt Bosley and Dennis Balthaser, meanwhile, demonstrated that the UFO community in America is still considered an open arena for their activities. Nor are the Mirage Men restricted to the USA. I spoke to a former employee of the Provost and Security Services, the UK's equivalent to AFOSI, who told me that they had run UFO-themed deception operations against a British ufologist in the 1990s. Perhaps Bosco Nedelcovic's claims of a worldwide UFO deception during the 1960s aren't so far-fetched.

Our journey had taught me that, even if we can't touch them, the UFOs are real. There's no escaping it. They are as old as humanity and they are here to stay. Like microwaves or solar flares, even if we can't see them, we know they are still out there.

Most UFOs may just be lights in the sky, but a light in the sky might change your life. In the right place at the right time a drifting Chinese lantern may have as much impact on a witness as a flying saucer piloted by Walt Bosley's Nazis from an alternate future.

So how should we really think about UFOs? Are they a problem of technology or of psychology; of semantics or ontology? Are they objects or events? Unidentified Flying Objects or Uncategorizable Fortean Occurrences? All we can say for certain is that they operate as much within the limitless domains of the imagination as they do in the skies above.

The stories that people tell about UFO events are just as important as the incidents themselves – and the two can be very different: in the remembering and the retelling, a Chinese lantern may eventually become a Nazi flying saucer from the Hollow Earth. And, as they grow more distant from their source, these stories expand and evolve, like ripples from a stone thrown into a pond, interfering and interacting with each other, and with us, the people who hear them. It's at this point that they become myths, and over time these myths shape our imaginations and the way that we experience the world. This is the true power of the UFO and that, ultimately, is what this book is about.

Myth-making is a human necessity. It's one of the things we do best. Myths are useful, they guide us and help us to make sense of the incomprehensible, those things and events too strange or complex for us to understand. They provide emotionally satisfying answers to difficult questions. How did we get here? Why is the world like it is? Why are we at war with people who were once our friends? Why did the World Trade Center collapse? Where do UFOs come from?

I have suggested that popular ideas about UFOs have been shaped and manipulated by disinformation specialists within America's intelligence apparatus: the Mirage Men. However, unlike Leon Davidson and Wernher von Braun, who allegedly believed that the US government was plotting a fake alien invasion, I don't think that there is any Grand Plan in the perpetuation of the UFO myth. The lore is perfectly capable of sustaining itself; it has a vibrant life of its own, supported by a complex patchwork of believers, promoters, seekers and charlatans, and nourished by the sightings of thousands of new witnesses every year. The lore does, however, provide a useful cover for certain clandestine operations, and is thus employed by the Mirage Men if and when it is expedient to do so.

The result is similar to what evolutionary biologists call a punctuated equilibrium: as the UFO myth has developed, it has been shaped by events such as the Maury Island incident, the Aztec crash, the abduction of Antonio Villas Boas, the Holloman landing, the Paul Bennewitz affair, the Majestic 12 documents and the Serpo papers. We can be sure that the Mirage Men were behind some of these punctuation marks, but a grand conspiracy is not only unnecessary, it is also unrealistic. As the FBI investigation into the MJ-12 papers shows, one arm of the intelligence community doesn't necessarily always know what another is doing, a situation that can only have increased the UFO's ability to insinuate itself into military and intelligence circles.

The UFO mythos is also highly complex: the range of experiences, events and phenomena that it encapsulates will never be explained by a single theory. The Mirage Men may be behind much of what we think about UFOs, but their forged documents and advanced technologies can't be everywhere at all times. Nor can they account for some of the more bizarre, complex and often very personal experiences that come under the ufological umbrella: the answers to these puzzles lie in the fields of psychology, meteorology and physics, not in espionage.

UFOs are a matter of belief, a faith whose core myths are taking on the shape of a religion; and, like any other faith, it deserves our respect, whether or not we share its tenets. People will always believe what they want to believe, what feels right to them, and for millions worldwide, extraterrestrials feel right. It's not a ridiculous thing to believe: there is water on Mars and on the Moon, and new Earth-like planets are found every year; it can only be a matter of time before we find something that we can all recognize as life elsewhere in the universe. But the discovery of life out there doesn't mean that life has been here and, even if it was capable of visiting us, it might be too wise or too uninterested to do so. Then again, as astrobiologists are now speculating, perhaps other life is already here in the form of microbes or, who knows, perhaps beneath the mountains of UFO reports lies a genuine case of otherworldly surveillance, or even contact. If so, I don't think the ufologists, or the Pentagon, have found it yet. And until such time as they do, the UFO community will continue to trace the same circles and loop the same strange loops that it has done since 1947.

Four years after being thrust into the ufological limelight, Bill Ryan is still there. He has left Serpo behind and now runs Project Camelot with Kerry Cassidy, whom he met at the Laughlin Convention in 2006. Since then Bill and Kerry have dedicated themselves to exposing the hidden truth of our interaction with extraterrestrial intelligences. What began with the Serpo exchange now involves cyborg soldiers, humanoid aliens on Earth, CIA bases on Mars, deadly ET viruses, teleportation and time-travel. Whether or not any of this new information is courtesy of AFOSI we may never know.

If an intelligence agency had created Bill as a PR man for the UFO myth, they couldn't have done a better job. Now, I don't think for one moment that they did – they didn't need to – but I do suspect that he was given a firm but gentle nudge in the right direction every now and then. Whether or not it was Rick Doty's role to do this, why, and for whom, remains unclear. Nobody knows, except perhaps Rick and whoever he may have been working with. And they're not talking, yet. Serpo, meanwhile is dead, or at least sleeping; Victor Martinez's email list carries more pictures of cute animals and swimsuit models than it does UFOs.

John and I are still in touch with Rick. Occasionally we talk crop circles, and every now and then he'll accuse me of being an MI6 agent, perhaps hoping that one day I'll crack and admit it. Just as this book was in its final stages, Rick emailed me after a long silence and restated his position on Serpo – 'the information came from a government employee working in the DC area' – and the wider UFO story:

Mark, no one currently within the UFO community knows the truth, nor do they have any first-hand information. They are relating second- or third-hand stories, or information from false documents created by false people . . . The most effective disinformation programme in US history came from the UFO community themselves. Why would the USG have to run any disinformation operations against them?

The truth is out there, you just won't believe it. US Intelligence is no longer involved in the UFO game. They were, at times, from about 1952 to 1985 . . . There was a fine line between promoting UFO sightings and protecting classified military projects. Most of our operations in this area were to protect classified projects from the Soviets. From 1952 we used various test sites in the southwest to experiment with classified aircraft (U-2, SR-71, F-117, and other craft that never made the market). If one of these planes was flying and a layperson saw it, we would try to convince them that what they saw was a UFO.

After the closure of Project Bluebook in 1969, the investigation of UFOs became an internal intelligence service investigation that was to be held classified. Why? Most intelligent people can figure this out quite easily. Because the US Intelligence Community knew what the UFOs were and were trying to keep a track on them. Common sense, Watson!!!!! We knew that some of the sightings were of craft that were intelligently controlled, and not by anything of this planet. We gathered intelligence, photographed them and turned the information over to analysts. Roswell was real. We obtained great technology from Roswell, but we returned the bodies in 1964 and established limited contact with the aliens from that point.

Each time I hear from Rick I wonder who he really is, and why he believes what he claims to believe. Did he really see something that was not of this Earth? Did something truly strange happen to him, or was something done to him to make think that it did? Or is he delusional, deceiving himself as much as he deceives others – one of Kit Green's paraphreniacs?

On a bad day I imagine Rick to be a ufological analogue of those fake Vietnam War veterans who wear medals that aren't theirs to curry respect and favour. But I think the truth is more complex than that. Rick's world – the entire intelligence world – is the trickster's domain, a world within our own, one without rules and that slides constantly between fantasy and reality. It's where the real UFOs are hidden.

The UFOs are tricksters, as are Rick and all the others like him. In traditional cultures it was the trickster's role not just to deceive, but to drive invention and science, just as UFOs always appear to possess technologies that lie around the corner from our own. Tricksters taught the spider to make her web and humans to make nets, traps and hooks; it was also common for tricksters to become caught in their own traps. Does Rick really believe that the ETs have been here? All I can say is that I hope so. But, whether or not Rick believes, millions of others do, and some of them hold positions of considerable influence.

John Podesta was Bill Clinton's Chief of Staff and managed Barack Obama's 2009 transition into the White House. A respected Washington politico, Podesta makes no secret of his passion for UFOs. In 2001 he's said to have had an *X Files*-themed fiftieth birthday party, hosted by the Clintons, who performed a Mulder and Scully routine as part of the entertainment. While Podesta remained circumspect on the UFO issue during his White House tenure, in 2002 he served as front man for the Coalition for Freedom of Information, a Si-Fi Channel-sponsored attempt to pry UFO documents from the government. 'It is time for the government to declassify records that are more than twenty-five years old and to provide scientists with data that will assist in determining the real nature of this phenomenon,' he told journalists at the project's launch.

Bill Richardson, Governor of New Mexico and a short-lived Democrat presidential nominee, served as Energy Secretary and UN ambassador under Clinton. He was chosen to be the Obama administration's Secretary of Commerce in 2009, but was forced to drop the role pending a financial investigation from which he was soon exonerated. In 2004 Richardson wrote the foreword to *The Roswell Dig Diaries*, a book spin-off from another Sci-Fi Channel event, an archaeological dig at one of the purported UFO crash sites. The dig found nothing of note; nevertheless Richardson wrote: 'The mystery surrounding this crash has never been adequately explained – not by independent investigators, and not by the U.S. Government . . . it would help everyone if the U.S. Government disclosed everything it knows. The American people can handle the truth – no matter how bizarre or mundane.'

How far could these beliefs go? George Bush Jr was able to tell the world that he received direction from God. Would America accept a president whose advice came from extraterrestrials? In the 1970s and 1980s Jimmy Carter and Ronald Reagan spoke publicly about their own UFO sightings and, while we don't know what they really knew or believed about the subject, we do know that Reagan raised the ET question more than once in a political setting.

The UFO issue is unlikely to make or break a president, or a potential president, but it might provide a useful weapon for their enemies. During the 2008 presidential campaign Democratic nominee Dennis Kucinich was outed as a UFO experiencer by his friend the actress Shirley Maclaine. The story of their shared encounter, dating back to 1982, provoked a media firestorm, with Kucinich pilloried by the pundits and forced into an awkward corner during a televised Q&A. While Kucinich was never going to be a presidential front runner, the UFO story was a damaging blow to his image as a serious candidate. It's unlikely that any future candidate will mention the UFO subject again, at least without a compelling public call to do so.

That call does exist, and it may be getting louder. In July 1999, 50,000 copies of the popular French weekly tabloid magazine *VSD* contained a ninety-page supplement entitled *UFOs and Defence, What Must We Prepare For*, put together by an organization called COMETA, the Committee for In-depth Studies. COMETA's members included a French admiral, an Air Force general and the former head of France's space programme. After stating that the UFO phenomenon can only be extraterrestrial in origin, the report identified the US as the nation most likely to have recovered debris from an alien craft, and to have made contact with the ETs. It concluded by recommending that France put diplomatic pressure on the US to reveal what it knows about the UFO subject and share the ET technologies at its disposal. The real purpose behind the COMETA report remains obscure, as do its true backers,[2] but its impact was felt internationally and helped to boost the credibility of the global extraterrestrial lobby.

Over the past decade an international movement under the umbrella term 'Disclosure' has generated a grass-roots push to force the world's governments, and the US government in particular, to tell the truth about UFOs. Disclosure's advocates claim that what they want is truth, but what they actually want is to have their existing beliefs about UFOs and extraterrestrials confirmed by a government they already distrust on the issue. The irony is exquisite. For six decades US government organizations have been trying to convince UFO witnesses that the aliens are real, but the Disclosure movement will not be satisfied with anything less than a formal presidential declaration, though I suspect that, even if they got it, they would still ask for more. So what do they want? DNA? Rick's Yellow Book? Free energy? A flying saucer? If anyone can provide it then the Mirage Men can. Except, of course, they won't; they don't need to: Disclosure is doing their work for them.

While the Disclosure movement's focus is on America, other nations have come clean and exposed themselves to the UFO lobby. Canada, Brazil, France and Britain have gone some way towards making their UFO files public, for the most part demonstrating that their governments were just as curious, and just as perplexed, as the rest of us. At about the same time that COMETA's report was published, a secret Ministry of Defence investigation, Project Condign, reached conclusions identical to those of many sensible ufologists: that beyond the usual misidentifications of UFOs there are a few genuine unknowns out there, most of them comprising 'black' aircraft and unusual natural phenomena. But conclusions such as these will never satisfy the Disclosure lobby. What they want to be told is that their myths are truths, even though many of these myths were created by disinformation specialists. So are some of those specialists working within the Disclosure movement, or at least watching it closely? Are there also agents of foreign intelligence services in its midst, calling on America to show us its technological crown jewels? I would guess that the answer to both questions is yes.

So what *is* up there? I've identified some of the aircraft and technologies that may have been mistaken for UFOs, or made to look like them, and there will have been others – some of them probably still flying today. It's also possible that real flying saucers were deployed by the Air Force or the Navy at one time; maybe they're still up there, or maybe they were superseded as a useful technology years ago. There's also evidence that military UFO reports of genuine significance, such as Robert Friend's 1955 alleged flying saucer sighting, were handled at a higher level than Project Blue Book; the compartmentalized nature of national security since the Second World War means that the military is constantly hiding secrets from itself.

The Pentagon is certainly hiding *something*. A new hangar sprung up at Area 51 in 2008, the largest yet built on the base, while the US Defense Department's proposed black budget for 2010 has leapt to $50 billion – that's $18 billion dollars more than for 2008 and more than Britain's entire military spend for the same year; $16 billion of that, more than twice as much as a decade ago, is dedicated to secret Air Force projects that might include the magical bird seen by Walt Bosley, or even my silver spheres.

Today's black budgets will produce tomorrow's UFOs. Technology has always advanced at the luminous edge of imagination – from Jules Verne's airships to today's laser weapons and cloaking devices – but the time may have come when scientific realities exceed the fantasies of science fiction. It might already be impossible for us to to distinguish human technologies from those that seem too-advanced-to-be-human, in fact that moment may have passed decades ago. We don't know; and that's exactly how the Mirage Men want to keep the situation.

We can be sure, however, that one day we will leave the Earth and become the extraterrestrials we have always dreamed of. The Disclosure movement talks about a US Space Navy: an armada of American militarized spacecraft encircling the Earth. This is what the British computer hacker Gary McKinnon claims to have found when accessing the Pentagon's computer network in 2001.

It's unlikely that McKinnon, by his own admission a fairly clueless hacker, was able to penetrate so deeply into the mainframes of the most sophisticated military on Earth. More probable is that he fell into a honey trap, one that provided him with exactly the information he was looking for; the kind of militaristic sci-fi fantasy that appeals to the hacker mentality and can only improve the Pentagon's technological profile on the international stage. Another British hacker, Matthew Bevan, was caught inside a US Air Force mainframe in 1996. He was also looking for UFO information and, like McKinnon, found it, though he came to believe that he had fallen victim to an extraterrestrial entrapment, a virtual-world creation of the Mirage Men.

McKinnon's Space Navy may one day become a reality, but for now it is a literalized metaphor for the real war in space, an intelligence war currently conducted by satellites. Even a space flotilla would cost a lot more than the $16 billion currently handed out to secret US Air Force projects – the six Apollo Moon landings of 1969–72 are estimated to have cost $145 billion in 2008 dollars. With the Space Shuttle currently on blocks, unless the Air Force is doing some thrifty saving, we're still a long way from *Battlestar Galactica*.

The real UFOs are imaginary weapons for psychological wars. In the right mindset, a UFO encounter can become a moment of ontological catastrophe, in which the boundaries that separate reality and fantasy, the forces that generate cause and effect, break down. The divisions between quotidian reality, our memories, dreams and imagination are not as sharply delineated as we like to think, and UFOs glide seamlessly between all these states of mind; they are at once real and unreal – trickster technologies, liminal engines. Like the intelligence agencies themselves, UFOs are immune to absolute truth and shield themselves with maybes. They are a classic source of uncertainty, at once neither/nor and both/and. It's tempting to label theirs a plasma state – like the luminous, fourth state of matter that is neither liquid nor solid and may account for a number of UFO reports.

UFOs are the ultimate tools of shock and awe, perfect weapons of mass deception. To understand how they are deployed, to see how they go to work in our hearts and minds, is to grasp the reality of the world we live in, saturated by media, bombarded by truths, half-truths, falsehoods and myths. It is the world of Weapons of Mass Destruction and the 9/11 Truthers; of Abu Ghraib and MMR; of MK-ULTRA and *Project Beta* – shifting forms occasionally illuminated and thrown into contrast by the dazzling lights of the UFOs.

The Mirage Men want us to believe simultaneously in both extremes of possibility: that the UFOs and their occupants are real, and that they have never existed at all. In the longer term, US Air Force policy seems to have been to encourage those who were already beyond discouragement while dissuading the professionals – the military men, the scientists, the engineers, the astronomers – who might see through their set-dressing to the smoke and mirrors beyond. As a UFO occupant told Nebraska patrolman Herbert Schirmer in 1967: 'We want you to believe in us, but not too much.'

Of course, the UFO will continue to exist without the Mirage Men. We need them. They awaken us to our roles as global citizens, they remind us of our ability to transcend national boundaries and become children of Earth. And eventually, one day, like Adamski's Orthon, we will become children of the Universe. UFOs, ETs and all the denizens of the otherworld are a part of our reality and a part of us; they are the psychic burnoff of an over-rationalized, over-mechanized society. The truth is that we are not rational beings and we will always find magic in our lives; and if we can't find it, we will create it. We don't need the Mirage Men to do this for us.

But what if there really is something more? What if I've just deceived myself into believing another conspiracy theory, my own sick-think? What if even a fraction of what Rick Doty says is real? What if Kit Green's core story contains a kernel of truth? If the ETs exist and they are here, how much could the Mirage Men really know about them? Are the Mirage Men and our alien visitors playing intelligence games with each other?

If these others are already a part of our world, then they haven't changed anything about it, or about us; and no beings, no matter how technologically advanced, can live up to the expectations that we would place upon them. Perhaps the aliens don't want to be our gods; perhaps it's not us that the Mirage Men are protecting, but them.

Somewhere between these two extremes of deception and communion, there might lie a strange and wonderful truth, and a future still waiting to happen.

But until then, the UFO must remain the domain of the Mirage Men.

NOTES

CHAPTER ONE: INTO THE FRINGE

1. David Clarke and Andy Roberts, *Out of the Shadows* (Piatkus, 2002).
2. Huyghe, Patrick, *Swamp Gas Times: My Two Decades on the UFO Beat* (Paraview Press, 2001).

CHAPTER TWO: THE COMING OF THE SAUCERS

1. For the inside story of the crop circle phenomenon, see *Round in Circles* by Jim Schabel (Penguin, 1994) and *The Field Guide* by Rob Irving and John Lundberg (Strange Attractor Press, 2006).
2. 'Court Rules Spy Book Extracts Can be Published', *Independent*, 26 January 2001.

CHAPTER THREE: UFO 101

1. Scenario described in *Deception* by Edward Jay Epstein (Simon & Schuster, 1989).
2. Jerome Clark, *The UFO Book* (Visible Ink, 1998). Could this be a twentieth-century prank?
3. Ibid. It's possible that this account may be the work of a twentieth-century prankster, of course.
4. David Clarke and Andy Roberts, *Out of the Shadows* (Piatkus, 2002).
5. A contraction of Detrimental Robot.
6. Fred Lee Crisman was no stranger to *Amazing Stories*, having had two letters published in its pages. In June 1946 he, like several other readers, had written in to describe an encounter with Richard Shaver's Dero in a cave in Burma. The second letter, published in May 1947, recalled another dramatic Dero incident, this time in Alaska, during which his companion was stuck and killed by a ray beam.
7. Arnold's friend E. J. Smith wondered where Ray Palmer found the $200 to give Arnold, or where Johnson got the money to cover his costs. It's been questioned as to whether or not that the US Army might have fronted the cash, though this needn't necessarily be the case: *Amazing Stories* was selling very well and, as a modest-sized newspaper, the *Statesman* also probably had money. However, a US Army sweetener can't be ruled out.
8. Kenneth Arnold and Ray Palmer, *The Coming of the Saucers* (privately published, 1952).

9. Chief spymaster Victor Perlo headed the aviation section, which might explain the interest in Arnold's announcements about atomic aircraft. As we'll learn later, there was a secret atomic aircraft project, though it was nowhere near take-off in 1947.
10. On which Oliver Stone's film *JFK* was based.
11. Via the FBI website, <fbi.gov>.
12. Via the NICAP website, <nicap.org>.
13. In a recently unearthed example, a list of rocket launches mentions a V-2 crashing near Roswell on 4 July, noting that it carried an unspecified 'biological payload', one of several to do so (see <www.rocketservices.co.uk/spacelists>).
14. See James Carrion, 'New Avenues for UFO Research', *MUFON Symposium Proceedings 2009*.
15. It was a theme that would be addressed by Ronald Reagan, also at a United Nations speech, forty years later: 'I occasionally think how quickly our differences worldwide would vanish if we were facing an alien threat from outside of this world . . . And yet I ask – is not an alien force already among us?' Reagan is said to have alluded to the idea repeatedly in discussions with the Soviets during the early years of *glasnost*, referring to it as his 'fantasy'. Reagan is also alleged to have been profoundly moved by a White House screening of Steven Spielberg's *ET* in June 1982 and to have told the director, 'You know, there aren't six people in this room who know how true this really is.' This odd comment has passed into UFO lore and its meaning can only be guessed at. I wrote to Spielberg asking whether this story was true, but my request got no further than his assistant.

CHAPTER FOUR: LIFT-OFF

1. Zeta Reticuli was made famous in UFO lore following the alleged 1961 abduction by aliens of mixed-race couple Betty and Barney Hill, which spawned a successful book, a TV film and launched thousands of similar stories. Under hypnosis by psychologist Benjamin Simon, Betty Hill, who died in 2004, believed that she was shown a map of her abductors' home star system. When Hill's abstract drawing of the map was published in John Fuller's 1966 book about the case, *The Interrupted Journey*, Marjory Fish, a schoolteacher from Ohio, was so inspired that she spent several years building a three-dimensional model of it – a work of cosmic devotional art. Thus Fish was able to identify the dual star system of Zeta Reticuli as the home planet of our alien visitors.
2. *Wired* online, 'Cybercrime Supersite "DarkMarket" was FBI Sting, Documents Confirm', 13 October 2008.

CHAPTER FIVE: CONFERENCE OF THE BIRDS

1. On 24 July 1948, Eastern Airlines Flight 571/23 reported a near miss with a rocket-like object over Virginia. The sighting was treated with the utmost seriousness by the US Air Force, who were developing rockets of their own. Could somebody else be flying rockets over the US? As the Air Force were aware, the Navy had launched a V-2 from the USS *Midway* in September 1947, and that the rocket might be theirs was almost as alarming a possibility as its being Russian.
2. Douglas Aircraft Inc., *Preliminary Design of an Experimental World-Circling Spaceship* (RAND, 1946).

3. Some supporters of the ET cover-up theory interpret these events quite differently. For them Forrestal's death was a state execution designed to prevent him from going public with what he knew about the Roswell UFO crash.
4. For a detailed account of this feud and its impact on UFO history, see Robert P. Horstemeier, 'Flying Saucers are Real! The US Navy, Unidentified Flying Objects, and the National Security State', *Socialism and Democracy*, Volume 20, Issue 3, November 2006.
5. Curtis Peebles, *Watch the Skies!* (Smithsonian Institution Press, 1994).
6. Nick Redfern, 'Incident at Aztec', *Fortean Times*, March 2004.
7. Karl T. Pflock, 'What's Really Behind the Flying Saucers? A New Twist on Aztec', *The Anomalist*, Spring 2000.
8. William Steinman and Wendelle Stevens, *UFO Crash at Aztec: A Well-kept Secret* (UFO Photo Archives, 1986).
9. John Marks and Victor Marchetti, *The CIA and the Cult of Intelligence* (Alfred A. Knopf, 1974). The book was considered so potentially damaging that 168 sections, including whole pages, were deleted by the CIA before its publication could be authorized. Marchetti resigned from the CIA in 1969. By the end of his fourteen-year career he had become special assistant to CIA Director Richard Helms.
10. Frances Stonor Saunders, *Who Paid the Piper? CIA and the Cultural Cold War* (Granta, 1999).
11. Ibid.
12. Ibid.
13. Ibid.
14. Carl Bernstein, 'The CIA and the Media', *Rolling Stone*, 20 October 1977.

CHAPTER SIX: WASHINGTON VERSUS THE FLYING SAUCERS

1. CIA memo ER-3-2809, 10 February 1952.
2. Jerome Clark, *The UFO Book* (Visible Ink, 1998).
3. Ibid.
4. Ruppelt claims that the temperature inversion was his own suggestion, one he had made offhand without serious consideration of the situation.
5. Churchill was a little slow off the mark here: a secret British MOD Directorate of Scientific Intelligence study had concluded the previous year that flying saucers were not a matter of defence concern. See Clarke and Roberts, *Out of the Shadows*.
6. Hadley Cantril, *The Invasion from Mars* (Princeton University Press, 1940).
7. Peebles, *Watch the Skies!*
8. *Washington Post*, 28 July 1952.
9. Leon Davidson, 'ECM+CIA=UFO', *Saucer News*, February/March 1959.
10. Quoted in ibid.
11. Ibid.
12. 'US Airman Milton Torres Told to Shoot Down UFO When Based at RAF Manston', *The Times*, 20 October 2008.
13. Eugene Poteat, 'Some Beginnings of Information Warfare, Stealth, Countermeasures, and ELINT, 1960–1975', *Studies in Intelligence*, Volume 42, No. 1, 1998.

14. *New York Times*, 30 July 1952.
15. Edward J. Ruppelt, *The Report on Unidentified Flying Objects* (Doubleday, 1956).
16. Ibid.
17. The document was discovered by astronomer and UFO researcher Jacques Vallée in 1967, though it wasn't made public until the publication of Vallée's diaries in 1992.
18. Leon Davidson, *The CIA and the Saucers* (Saucerian, 1976).

CHAPTER SEVEN: PIONEERS OF SPACE

1. Leon Davidson, *The CIA and the Saucers* (Saucerian, 1976).
2. Nick Redfern, *On the Trail of the Saucer Spies* (Anomalist Books, 2006).
3. Orfeo Angelucci, *The Secret of the Saucers* (privately published, 1955).
4. Charles Bowen (ed.), *The Humanoids* (Futura, 1974).
5. Magnesium began to be used in the airframes of American aircraft in the 1950s.
6. Probably a reference to an alleged saucer crash on the island of Spitzbergen in Norway in 1952. The story first appeared in a German newspaper, *Berliner Volksblatt*, on 9 July 1952 and was quickly circulated internationally. A National Security Agency airgram about flying saucers stories in the Russian press, dated 20 February 1968, makes reference to the Spitzbergen event, which is circled and labelled 'Plant'.
7. Bowen, *The Humanoids*.
8. Ibid.
9. It's unlikely that a CIA project would so clearly reflect its aims. Nedelcovic also mentioned Operation Exeter, conducted simultaneously in Exeter, New Hampshire, USA and Exeter, Devon, UK in 1965. The UK incident involved a humanoid ET who identified himself as 'Yamski' and took place on 24 April 1965, the day after George Adamski's death. Was somebody having fun with this one?
10. This may seem crowded, but the Sikorsky H-34 Choctaw, or S-58, in use from 1954, could carry sixteen people.
11. Speaking in 1978, Nedelcovic identified the drug as Lorazepam, a powerful benzodiazepine tranquillizer and muscle relaxant with sedative and hypnotic effects. Lorazepam was introduced in 1977, but the benzodiazepine family was discovered in 1954.
12. The experiments were widely reported in the media. See for example Kathryn A. Braun, Rhiannon Ellis and Elizabeth F. Loftus, 'Make My Memory: How Advertising Can Change Our Memories of the Past', *Psychology & Marketing*, January 2002.
13. The notion of 'brainwashing', a word coined by CIA-friendly journalist Edward Hunter, was intended to explain why some American soldiers had expressed Communist sympathies after being captured by the North during the Korean War.
14. This is not to suggest that all subsequent 'alien abductions' were the result of psychological warfare experiments on innocent civilians. The Villas Boas case and the complex 1961 abduction account of Betty and Barney Hill served as historical precedents and authorities for these experiences, the vast majority of which have their roots in sleep disorders and idiosyncrasies of memory.

CHAPTER EIGHT: AMONG THE UFOLOGISTS

1. For reasons of his own, Moore declined to speak to John and myself – the only person not

employed by the US government to do so. The account here is taken from the MUFON video, and from Greg Bishop's *Project Beta* (Paraview, 2005).

2. Jerry Miller had been an investigator for Project Blue Book and was considered to be highly knowledgable about UFOs.

3. Jerome Clark, *UFOs in the 1980s (The UFO Encyclopedia, Volume 1)* (Apogee Books, 1990).

4. Greg Bishop, *Project Beta* (Paraview, 2005).

5. For example the section bracketed with (S/WINTEL) should instead read S/WNINTEL, for 'Warning Notice Intelligence Sources & Methods Involved'. From <www.cufon.org/cufon/foia_004.htm>.

6. The USCGS was submerged into the National Oceanic and Atmospheric Administration. Barry Greenwood and Brad Sparks; *The Secret Pratt Tapes and the Origins of MJ-12*', *MUFON Symposium Proceedings, 2007.*

7. Greg Bishop, *Project Beta.*

8. Ibid.

9. Moore has always denied receiving payment for his AFOSI work, and I have seen no evidence to the contrary.

10. See for example <www.nicap.org/foia_003.htm>.

11. Taken from the video recording of Moore's presentation.

CHAPTER NINE: RICK

1. Christa Tilton, *The Bennewitz Papers* (Crux Publications, 1991).

2. In a January 1988 telephone inteview with sceptical UFO investigator Philip Klass, Doty stated that Ed Doty was his father and 'was an investigator for Blue Book, he was at Holloman from 1962 to 1966, four years. He was involved in the Lonnie Zamora case' (the landing of a UFO in Socorro, New Mexico in1964). From this and other discrepancies over the years it would seem that Doty may have extended his disinformation training into his personal life.

3. The book is a strange pot pourri of alleged, mostly second-hand UFO and ET tales from military and intelligence insiders. It lists Doty as co-author and appeared in 2006, seemingly riding the coat tails of the Serpo furore.

4. Founded in 1956, key NICAP members included Donald Keyhoe, Admiral Roscoe Hillenkoetter, another Navy man who also happened to have been the first director of the CIA and Colonel Joseph Bryan III, the CIA's first Chief of Political and Psychological Warfare. Understandably some have wondered if NICAP was set up as a CIA conduit into the UFO community.

5. Terry Hansen, *The Missing Times* (Xlibris, 2000).

6. The recipient of the original letter is unknown, but it is dated 3 March 1989 and was later distributed on the Internet.

7. Crystal-based technologies occur in the early UFO literature, such as contactee Orfeo Angelucci's *The Secret of the Saucers* (1955). Crystals had, of course, been considered sources of mystical knowledge for centuries before UFOs appeared in our skies.

CHAPTER TEN: BEEF BUGS AND THE ALIEN UNDERGROUND

1. In April 2009 the hotel hosted the first ever Dulce UFO conference. More than a hundred people showed up, when the organizer had expected only thirty. Late arrivals were turned away.
2. Roberta Donovan and Keith Wolverton, *Mystery Stalks the Prairie* (THAR Institute, 1976).
3. Bob Pratt hinted as much to journalist Terry Hansen, author of *The Missing Times*.
4. Colm Kelleher, *Brains Trust* (Paraview, 2004).
5. It's a story that has never died. The alleged base at Dulce was the subject of a full-length History Channel documentary in 2009.

CHAPTER ELEVEN: OVER THE EDGE

1. Robert Collins and Richard Doty, *Exempt from Disclosure* (Peregrine Communications, 2006).
2. It's rumoured that NASA tried to persuade Spielberg not to release the film, fearing that it would increase pressure on them to open a new official UFO investigation.
3. AP, 26 November 1977.
4. Barry Greenwood and Brad Sparks, 'The Secret Pratt Tapes and the Origins of MJ-12', *MUFON Symposium Proceedings, 2007*. The authors also discuss the NASA issue in some detail.
5. This fact is omitted in the Moore/Doty/Bishop version of events.
6. *Air Force Policy Directive 10-7*, 'Operations, Information Operations', 6 September 2006.

CHAPTER TWELVE: SEEING IS BELIEVING

1. In the name of research I watched the first five seasons of *The X Files* looking for Rick and didn't find him. I was later able to interview the series' creators, Chris Carter and Frank Spotnitz, neither of whom knew of Rick, though this doesn't disprove his claims.
2. When Bob told us this story our cameraman, Grant Wakefield, announced that he had seen a very similar demonstration given to professional camera operators just a couple of years previously. A holographic 'Telepresence' technology for remote, real-time holographic business conferencing was introduced by CISCO in 2006.
3. Memo from Brigadier General C. H. Bolender, US Air Force, 20 October 1969, via <nicap.org/Bolender_Memo.htm>.
4. It was Lundahl who first spotted Soviet missile launchers on Cuba, instigating the 1962 missile crisis.
5. Jay Gourley, 'The Day the Navy Established Contact', *Second Look*, May 1979. Emenegger tried to get Lundahl to speak about the incident on camera, but he refused to do so.
6. Robert Emenegger, *UFOs: Past Present and Future* (Ballantine, 1973).
7. It's likely that AFOSI and/or Doty had a hand in these documents, the authenticity of which has now been called into question.
8. As of late 2009, the Socorro sighting, one of the most celebrated in UFO history, is looking likely to have been an elaborate hoax perpetrated on Zamora by students at the New Mexico Institute of Mining and Technology.

CHAPTER THIRTEEN: BAD INFORMATION

1. Ellic Howe, *The Black Game* (Futura, 1988).
2. This might lend credence to Bosco Nedelcovic's claims.
3. Howard Blum, *Out There* (Simon & Schuster, 1990).
4. Ibid.
5. Barry Greenwood and Brad Sparks, 'The Secret Pratt Tapes and the Origins of MJ-12', *MUFON Symposium Proceedings 2007*.
6. The duo would later become key sources for the Stealth model kit developed by the Testors Corporation; this was released in 1986, two years before the Air Force went public with their real aeroplane, and proved a major security and public relations disaster for the US Air Force.

CHAPTER FOURTEEN: FLYING SAUCERS ARE REAL!

1. The DED is busier than ever, its projected budget for 2010 leaping from $62.7 million to $105.7 million.
2. Gerald K. Haines, 'A Die-Hard Issue: CIA's Role in the Study of UFOs, 1947–90'. *Studies in Intelligence*, Semiannual Edition, No. 1, 1997. Intriguingly, footnote 90 of the essay reads: 'The CIA reportedly is also a member of an Incident Response Team to investigate UFO landings, if one should occur. This team has never met.'
3. While Haines's revelations satisfied many, the pro-ET wing of the UFO community weren't convinced. Mark Rodeghier of the Center for UFO Studies contacted Robert Friend, Blue Book's penultimate head from 1958 to 1963, who flatly denied Haines's claims about the U-2, although he admitted that Blue Book did maintain regular contact with the CIA and had, on occasion, concealed classified operations as UFOs. Knowing that Friend passed Bob Emenegger dodgy documents back in 1973, however, we might not want to take his word as gospel.
4. The Upshot-Knothole detonations exposed thousands of people across southern Nevada to potentially dangerous levels of radiation. Thousands of sheep died within weeks of the tests, and all the while the Atomic Energy Commission insisted that the explosions had been contained and the dangers of atomic fallout were minimal.
5. UFO researcher Nick Redfern proposes a human experiment variation of this idea as the reason for the Roswell cover-up in his book *Body Snatchers in the Desert* (Paraview, 2005).
6. As if to prove this point, on 14 September 2009 the *Sun* website carried a still of a US Air Force F-18 fighter jet being shadowed by a smaller 'UFO'. In reality it was a still from a Pentagon video showing the Boeing X-45 drone in action.
7. The name refers to a running joke about the violinist Yehudi Menuhin appearing on the Bob Hope radio show.
8. Zimmerman also patented an unusual flying platform, which was successfully flown in 1955.
9. Tony Buttler and Bill Rose, *Flying Saucer Aircraft*. (Midland, 2006).
10. Leon Davidson, *The CIA and the Saucers* (Saucerian, 1976).
11. See the works of Wilhelm Landig and Ernst Zundel.
12. Martin Kottmeyer, 'Missing Linke', *Magonia Monthly Supplement*, 21 October 2003.

CHAPTER FIFTEEN: THE SECRET WEAPON

1. C. G. Jung, *Flying Saucers: A Modern Myth of Things Seen in the Sky* (Routledge, Kegan & Paul 1959).
2. Although this sounds like a lot of money, it's worth comparing it with the $1.7 million given to Boeing in 1946 to build a single mock-up B-52 Stratofortress.
3. Leon Davidson, *The CIA and the Saucers* (Saucerian, 1976).
4. See the issue for July 1957. An electrogravitic, ion-propelled craft also featured on the August 1964 cover of *Popular Mechanics*. It's speculated that this technology has been put to use by the US Air Force, perhaps in the larger, slow-moving triangular UFOs seen in the past two decades.
5. After a close encounter with a diamond-shaped craft in Texas, Betty Cash and Vickie Landrum suffered what doctors identified as serious gamma-radiation burns. The duo took the US government to court aided by attorney Peter Gersten. In January 1983, as Gersten was preparing his case, he had a two-day meeting with Richard Doty at Kirtland AFB. Doty told Gersten about the alien presence on Earth and asked if he could join his organization, Citizens Against UFO Secrecy. Gersten refused Doty's request. Curiously, at the end of the meeting Doty admitted to Gersten that AFOSI was passing disinformation to the UFO community (see Barry Greenwood and Brad Sparks, 'The Secret Pratt Tapes'). In 1986, after a protracted legal battle, Cash, Landrum and Gersten lost their case against the government.
6. Allegedly a pen name of aviation journalist James Goodall. The very first mention of UFOs at Area 51 seems to have been in the *Mutual UFO Network Journal* for September 1980. A military witness describes seeing a silent, twenty to thirty-foot-diameter saucer being flown as part of Project Redlight.
7. R. V. Jones, *Most Secret War: British Scientific Intelligence 1939–1945* (Hamish Hamilton, 1978).

CHAPTER SIXTEEN: A MATTER OF POLICY

1. Victor Marchetti, and John Marks, *The CIA and the Cult of Intelligence* (Alfred A. Knopf, 1974).

CHAPTER EIGHTEEN: WEAPONS OF MASS DECEPTION

1. On Kerrang! Radio, a British heavy-metal channel.
2. A French physicist, Jean Pierre Petit, is thought to have been one of the key figures behind COMETA. Petit believes that the US and Russian militaries have been experimenting with magnetohydrodynamics and the coanda effect in flying saucer designs since the 1950s, and that the inspiration came from downed ET craft.

BIBLIOGRAPHY AND FURTHER READING

BOOKS QUOTED AND REFERENCED IN THE TEXT

Agee, Philip, *Inside the Company: CIA Diary* (Penguin, 1975)

Angelucci, Orfeo, *The Secret of the Saucers* (Privately published, 1955)

Arnold, Kenneth and Ray Palmer, *The Coming of The Saucers* (Privately published, 1952)

Berlitz, Charles and William Moore, *The Roswell Incident* (Granada, 1982)

Bishop, Greg, *Project Beta: The Story of Paul Bennewitz, National Security, and the Creation of a Modern UFO Myth* (Paraview, 2005)

Blum, Howard, *Out There: The Government's Secret Quest for Extraterrestrials* (Simon & Schuster, 1990)

Bowen, Charles (ed.), *The Humanoids* (Futura, 1974)

Buttler, Tony and Bill Rose, *Flying Saucer Aircraft* (Midland, 2006)

Clark, Jerome, *The UFO Book* (Visible Ink, 1998)

Clarke, David and Andy Roberts, *Out of the Shadows: UFOs, the Establishment and Official Cover Up* (Piatkus, 2002)

Collins, Robert and Richard Doty, *Exempt from Disclosure* (Peregrine Communications, 2006)

Davidson, Leon, *Flying Saucers: An Analysis of the Air Force Project Blue Book Special Report 14 including The CIA and the Saucers* (Saucerian, 1976)

Donovan, Roberta and Keith Wolverton, *Mystery Stalks the Prairie* (THAR Institute, 1976)

Dewey, Steve and John Ries, *In Alien Heat: The Warminster Mystery Revisited* (Anomalist Books, 2006)

Emenegger, Robert, *UFOs: Past Present and Future* (Ballantine, 1973)

Epstein, Edward Jay, *Deception: The Invisible War Between the KGB and the CIA* (Simon & Schuster, 1989)

Festinger, L., H. W. Riecken and S. Schachter, *When Prophecy Fails* (University of Minneapolis Press, 1956)

Fuller, John, *Incident at Exeter* (Berkley, 1966)

Hansen, Terry, *The Missing Times* (Xlibris, 2000)

Howe, Ellic, *The Black Game: British Subversive Operations Against the Germans During the Second World War* (Futura, 1988)

Huyghe, Patrick, *Swamp Gas Times: My Two Decades on the UFO Beat* (Paraview, 2001)

Jones, R. V., *Most Secret War: British Scientific Intelligence 1939–1945* (Hamish Hamilton, 1978)

Jung, C. G. *Flying Saucers: A Modern Myth of Things Seen in the Sky* (Routledge, Kegan & Paul, 1959)

Keel, John A., *The Mothman Prophecies* (Saturday Review Press, 1975)

Kelleher, Colm, *Brains Trust: The Hidden Connection Between Mad Cow and Misdiagnosed Alzheimer's Disease* (Paraview, 2004)

Kraspedon, Dino, *My Contact with Flying Saucers* (Neville Spearman, 1959)

Marks, John and Victor Marchetti, *The CIA and the Cult of Intelligence* (Alfred A. Knopf, 1974)

Newman, Bernard, *The Flying Saucer* (Victor Gollancz, 1948)

Peebles, Curtis, *Watch the Skies! A Chronicle of the Flying Saucer Myth* (Smithsonian Institution Press, 1994)

Pflock, Karl and Peter Brookesmith (eds), *Encounters at Indian Head: The Betty and Barney Hill UFO Abduction Revisited* (Anomalist Books, 2007)

Redfern, Nick, *Body Snatchers in the Desert* (Paraview, 2005)

—, *On the Trail of the Saucer Spies* (Anomalist Books, 2006)

Ruppelt, Edward J., *The Report on Unidentified Flying Objects* (Doubleday, 1956)

Saunders, Frances Stonor, *Who Paid the Piper? CIA and the Cultural Cold War* (Granta, 1999)

Scully, Frank, *Behind the Flying Saucers* (Henry Holt & Co., 1950)

Snyder, Alvin A., *Warriors of Disinformation: American Propaganda, Soviet Lies, and the Winning of the Cold War* (Arcade Publishing, 1995)

Tilton, Christa, *The Bennewitz Papers* (Crux Publications, 1991)

Vallée, Jacques, *Forbidden Science: Journals 1957–1969 (North Atlantic Books, 1992)*

PAPERS QUOTED AND REFERENCED IN THE TEXT

Air Force Policy Directive 10-7, 'Operations, Information Operations'. 6 September 2006

Bernstein, Carl, 'The CIA and the Media', *Rolling Stone*, 20 October 1977

Braun, Kathryn A., Rhiannon Ellis and Elizabeth F. Loftus, 'Make My Memory: How Advertising Can Change Our Memories of the Past', *Psychology & Marketing*, January 2002

Carrion, James, 'New Avenues for UFO Research', *MUFON Symposium Proceedings 2009*

Davidson, Leer, 'ECM+CIA=UFO', *Saucer News*, Feb/March 1959.

Douglas Aircraft Inc., *Preliminary Design of an Experimental World-Circling Spaceship* (RAND, 1946)

Gourley, Jay, 'The Day the Navy Established Contact', *Second Look*, May 1979

Greenwood, Barry and Brad Sparks, 'The Secret Pratt Tapes and the Origins of MJ-12', *MUFON Symposium Proceedings 2007*

Haines, Gerald K., 'A Die-Hard Issue: CIA's Role in the Study of UFOs, 1947–90', *Studies in Intelligence*, Semiannual Edition, No. 1, 1997

Horstemeier, Robert P., 'Flying Saucers are Real! The US Navy, Unidentified Flying Objects, and the National Security State', *Socialism and Democracy*, Volume 20, Issue 3, November 2006

Kottmeyer, Martin, 'Missing Linke', *Magonia Monthly Supplement*, 21 October 2003

Pflock, Karl T., 'What's Really Behind the Flying Saucers? A New Twist on Aztec', *The Anomalist*, Spring 2000

Poteat, Eugene, 'Some Beginnings of Information Warfare, Stealth, Countermeasures, and ELINT, 1960–1975', *Studies in Intelligence*, Volume 42, No. 1, 1998

Redfern, Nick, 'Incident at Aztec', *Fortean Times*, March 2004

USEFUL FURTHER READING

Bamford, James, *Body of Secrets: Anatomy of the Ultra-Secret National Security Agency from the Cold War Through the Dawn of a New Century* (Arrow Books, 2002)

Blum, William, *Killing Hope: U.S. Military and CIA Interventions Since World War II,* (revised edition, Zed Books, 2003)

Cantril, Hadley, *The Invasion From Mars* (Princeton University Press, 1940)

Clarke, David and Andy Roberts, *Flying Saucerers: A Social History of UFOlogy* (Alternative Albion, 2007)

Dolan, Richard M., *UFOs and the National Security State* (Keyhole Publishing, 2000)

Fawcett, Larry and Barry Greenwood, *Clear Intent: The Government Coverup of the UFO Experience* (Prentice Hall, 1984)

Goodrick Clarke, Nicholas, *Black Sun: Aryan Cults, Esoteric Nazism and the Politics of Identity* (New York University Press, 2002)

Hambling, David, *Weapons Grade: Revealing the Links Between Modern Warfare and Our High-Tech World* (Constable, 2005)

Hansen, George P., *The Trickster and the Paranormal* (Xlibris, 2001)

Hill, Paul R., *Unconventional Flying Objects: A Scientific Analysis* (Hampton Roads Publishing Co., 1995)

Hollings, Ken, *Welcome to Mars: Fantasies of Science in the American Century 1947–1959* (Strange Attractor Press, 2008)

Hynek, J. Allen, *The UFO Experience: A Scientific Inquiry* (Ballantine Books, 1972)

Keel, John A., *UFOs: Operation Trojan Horse* (Abacus, 1973)

Maccabee, Bruce, *The Secret History of the Government's Cover-Up* (Llewellyn Publications, 2000)

Matthews, Tim, *UFO Revelation: The Secret Technology Exposed?* (Blandford, 1999)

Patton, Phil, *Travels in Dreamland: The Secret History of Area 51* (Gollancz, 1998)

Rankin, Nicholas, *Churchill's Wizards: The British Genius for Deception, 1914–1945* (Faber & Faber, 2009)

Schuessler, John, *The Cash–Landrum UFO Incident* (GeoGraphics Printing Company, 1998)

Streatfield, Dominic, *Brainwash: The Secret History of Mind Control* (Hodder & Stoughton, 2006)

Thomas, Kenn, *Maury Island UFO* (IllumiNet, 1999)

Vallée, Jacques, *Messengers of Deception* (Ronin Publishing, 1979)

Vallée, Jacques, *Confrontations: A Scientist's Search for Alien Contact* (Ballantine Books, 1990)

Vallée, Jacques, *Dimensions: A Casebook of Alien Contact* (Ballantine Books, 1995)

Vallée, Jacques, *Revelations: Alien Contact and Human Deception* (Ballantine Books, 1992)

Ward, Bob, *Dr. Space: The Life of Wernher von Braun* (US Naval Institute Press, 2005)

ACKNOWLEDGEMENTS

First and foremost I'd like to thank John Lundberg, without whom none of this would have happened, and without whose grounding, insight, humour and drive it would have remained nothing more than a Bliss Cafe reverie.

Mirage Men was always conceived simultaneously as a film and a book. Each medium is capable of things that the other isn't, and I hope that ours will serve as ideal complements to each other. Big thanks, then, to our two film crews – Zillah Bowes and Emma Meaden, and Andrew Brown and Grant Wakefield – for all their hard work and professional input; to Barry Hale and Uzma Choudhry at Threshold Studios, Northampton for their time and assistance; to Bob and Margaret Emenegger for their warm Arkansas hospitality; to Jon Ronson for his early assistance; to George Duffield; and to Mary Burke and Barry Ryan at Warp Films for their initial enthusiasm for the project.

While writing a book is an uncompromisingly solitary affair, this one happened only with help from a number of people. Huge thanks are due to Hannah Westland and Rowan Routh at RCW for their faith, encouragement and support in getting *Mirage Men* off the ground; to Andreas Campomar and Leo Hollis at Constable & Robinson for taking on such an unusual proposition, for their patience in awaiting its arrival and to Leo for his sensible editorial suggestions; to Phil Baker, Louise Burton, Alex Butterworth, Ken Hollings, Mike Jay, Gary Lachman and S. F. Said for their advice along the way, and to LSD for putting up with me as I vanished into the wormhole.

My gratitude also to (almost) everyone in the UFO and intelligence communities who spoke to John and me while *Mirage Men* came to fruition. I'd particularly like to thank Greg Bishop for his time, information, assistance and friendship; Walt Bosley, Rick Doty and Kit Green for agreeing to speak to us about their work; Caryn Ascomb and Jack Sarfatti for many enjoyable conversations and for defending me against the forces of ufological fanaticism – sadly neither of their fascinating stories has made the final cut; Jacques Vallée for his encouragement and insight; Bill Knell and Rich Reynolds for supplying essential research materials; John Rimmer and John Harney at *Magonia* and everyone at *Fortean Times* for keeping the torch of informed and intelligent high strangeness aflame.

Finally, the UFO phenomenon is only as interesting as what's been said and written about it, so I'd like to express my appreciation for all those people – sceptics and believers, seekers and chancers – whose lives and works have made the UFO story as fascinating and entertaining as it is. In particular, I want to pay tribute to John Alva Keel, author, adventurer and two-fisted trickster, to whose memory *Mirage Men* is dedicated.

INDEX